RAE GOODELL

The
Visible
Scientists

LITTLE, BROWN AND COMPANY — BOSTON – TORONTO

FIRST EDITION

T05/77

The author acknowledges the reprinting of previously copyrighted materials as follows:

"Academe and I" by Isaac Asimov, from *The Magazine of Fantasy and
Science Fiction*, May, 1972. Copyright © 1972 by Mercury Press, Inc.

"The Prime of Life" by Isaac Asimov, from *The Magazine of Fantasy
and Science Fiction*, October, 1966. Copyright © 1966 by Mercury Press,
Inc.

"Is Quality of U. S. Population Declining?" an interview with William
Shockley from *U. S. News & World Report*, November 22, 1965.

"Faking It" by Arthur Herzog, from *Saturday Review of Society*,
Volume 1, April, 1973. Copyright © 1973 by Saturday Review/World,
Inc.

LIBRARY OF CONGRESS CATALOGING IN PUBLICATION DATA
Goodell, Rae.
    The visible scientists.
    Based on the author's thesis.
    Includes bibliographical references and index.
    1. Science — Social aspects — United States.
    2. Technology — Social aspects — United States.
    3. Scientists — United States. I. Title.
Q175.52.U5G66   509'.2'2   77-350
ISBN 0-316-32000-5

Designed by Susan Windheim
Published simultaneously in Canada
by Little, Brown & Company (Canada) Limited

PRINTED IN THE UNITED STATES OF AMERICA

To my mother,
Dorothy P. Whitten

Acknowledgments

I had been interviewing Margaret Mead for some time one afternoon in her study when her phone rang. Feeling that this made a good breaking point, I rose to go. Oh, no, she instructed me, I was to wait: a project of this magnitude deserved more of her time. For the cooperation and hospitality of the visible scientists and their colleagues, I am grateful. Although interviews were often "long and complicated," as B. F. Skinner put it, each of the visible scientists gave a personal demonstration of the generosity that makes this group attractive to journalists.

Many people deserve thanks for their assistance in the preparation of the dissertation on which this book is based, especially William L. Rivers, my thesis advisor. Throughout the long study, he provided step-by-step guidance and comma-by-comma editorial advice, as he has for so many fortunate students. At Stanford University William J. Paisley also played a special role in the conceptualization of the dissertation, particularly by suggesting the term "visible scientists." For help in early stages I would also like to thank Martin Perl, Hillier Krieghbaum, James Butler, Allan Mazur, and Richard Stephens.

Some of the research was funded by a doctoral dissertation grant from the National Science Foundation, and a travel grant from the Office of Communications Programs for the Public

Understanding of Science of the American Association for the Advancement of Science.

The comments of people who read the dissertation were very helpful in making book revisions; those who gave the manuscript particular attention (although they of course are not responsible for remaining errors) include, in addition to Rivers and Paisley, Edwin Parker, Donald Roberts, Nathan Maccoby, David Noble, Sally Hacker, Charles Weiner, and my very capable Little, Brown editor, Bill Phillips.

Most important, this book was a family venture. My mother, Dorothy P. Whitten, transformed barely readable pages into workable drafts and then to final form, saving me countless hours of work. At the same time, Charles Whitten provided invaluable assistance, for example by proofreading the manuscript. Ross Goodell, a computer engineer and a poet, gave the book both its practical support and its creative energy. Assuming financial responsibility, sharing domestic work, and offering stimulating ideas, he made the book his own. And Maia Goodell, age six, put it all in perspective.

Contents

The Visible Scientists

Introduction

A pale, balding, bespectacled professor in a white laboratory coat steps to the microphone, blinks uncomfortably at the bright camera lights, unfolds a prepared statement from his pocket, and reads it verbatim in a quavering voice. He is then subjected to polite questions from reporters, which he answers in muddy, technical terms. Occasionally he scribbles obscure equations on the blackboard. Mercifully, the session is short, and the scientist is allowed to-slip peacefully back to his beloved laboratory, where, among rows of gleaming test tubes and bubbling retorts, he can continue to pursue his life's work, his magnificent obsession.

Such has been our image of the scientist and how he might handle a rare and reluctantly granted press conference.* The scientist, as we picture him, is not a people person and certainly not a press person. He is probably not even interested in politics, with its hustle-bustle and lack of exactness. He leaves political questions and social amenities to those with more time and inclination. He feels, instead, a moral obligation and hedonistic passion to add his small increment to scientific knowledge.

---

* References for factual information and quotations are found in the Notes, pp. 208–233.

He lives in isolated splendor, consumed by his work, and dies in obscurity, understood only by his colleagues.

Our image scientist has one other striking characteristic: he is almost the opposite of the kind of scientist we actually see in the news media today. Like the advertising industry's version of the liberated woman, today's visible scientist has "come a long way." The liberation of women is sometimes depicted with "before" and "after" pictures. In the "before" scene, the lady is demure, lace from head to toe, half hidden by a parasol; "after," she is bold, brassy, bikinied. Our image of scientists is a "before" picture.

This is a book about the "after" scientists — the scientists who are visible to the general public today. What distinguishes them is not that they are better or less known than their predecessors, but that they are known for a different reason. Typically, in the past, the public has known an occasional discoverer: Isaac Newton, Albert Einstein, Jonas Salk. In nineteenth century America, many of the commonly known scientists were famous as inventors: Joseph Henry, Thomas Edison, Robert Fulton, Alexander Graham Bell. An exceptional scientist might also achieve a name as a lecturer or popularizer: Humphrey Davy, Thomas Henry Huxley, Michael Faraday. Science-watchers also knew the leaders of the scientific community, those who served as spokesmen for their fellows: Robert Millikan, for example, after World War I, Vannevar Bush after World War II.

Today's scientists become visible primarily not for discoveries, for popularizing, or for leading the scientific community, but for activities in the tumultuous world of politics and controversy. Aggressively taking advantage of the new communications media, they seek to influence people and policy on science-related subjects — overpopulation, drugs, genetic engineering, nuclear power, pollution, genetics and IQ, food shortages, energy shortages, arms control. Circumventing the traditional channels for influencing science policy, they take their message directly to the public. To succeed, they must be knowledgeable, articulate, dramatic, persistent, and sophisticated about press operations. Those who do succeed become known to the public not for their science but for their public involvement:

— *Paul Ehrlich:* He is out to sell the public on halting the population explosion, by combining persistence, a large following, and Madison Avenue techniques. "If they can sell flavored douches, for chrissake," he says, "we can sell anything."

— *Linus Pauling:* He is one of America's most brilliant and productive scientists. But laymen know him for a book on vitamin C, and if they're a little older, for his agitation for disarmament and peace.

— *Margaret Mead:* For almost fifty years she has been the people's anthropologist. Moving from sex customs to the generation gap to environment, she has translated anthropology into relevant and practical terms for parents, students, and policymakers.

— *B. F. Skinner:* He antagonizes psychologists and laymen alike, for his contention that man's behavior is determined by the environment. But his theories are as influential as they are controversial, both in psychology and lay society.

— *Carl Sagan:* He is the Pied Piper of astronomy, captivating youngsters and taxpayers with the wonders of the "cosmic overwhelm."

— *Barry Commoner:* The "Paul Revere of ecology," he turns his back on the scientific and political establishment because the ultimate decision-makers are the public.

— *William Shockley:* He views himself as a Saviour, and is viewed by most of his peers as a Satan. *He* believes he has only the few years remaining in his life to alert the country to impending genetic suicide. *They* believe he is fanning the fires of racism, with his views that blacks are genetically inferior to whites. He cultivates the friendship of right wing newspaper editors, and welcomes even adverse publicity, anything to get his ideas into print.

These are the scientists the public knows today, to the extent it knows scientists at all. They are clearly not the good old-fashioned famous scientists, known primarily for research discoveries; it is a rare layman who knows that Linus Pauling won the Nobel Prize for elucidating the nature of the chemical bond, or that Paul Ehrlich's research area is lepidoptery (butterflies). Nor are they popularizers in the usual sense of the word; what they "popularize" is science-related policy issues, not science, and they often make science *un*popular in the process. Nor are they the leaders of the scientific community; on the contrary, they are typically outsiders, sometimes even outcasts among established scientists, speaking from personal conscience, not group consensus.

That a change should be taking place in the visibility of scientists is not surprising. Like politicians, actors, or football players, scientists gain visibility largely through the communications media, and the media have undergone revolutionary change in the past few decades. Most of us heard it first from Marshall McLuhan: "The medium, or process, of our time — electric technology — is reshaping and restructuring patterns of social interdependence and every aspect of our personal life. . . . Everything is changing — you, your family, your neighborhood, your education, your job, your government, your relation to 'the others.' And they're changing dramatically."

Concurrently, the uneasy relationship between science and the public has been changing, as technological ills have increasingly plagued society. These changes in turn have put pressure on science to update its antiquated concepts of how much to tell the public, when, and how. "It's time," warned Jules Bergman, ABC's science editor, "for the scientific community to awaken to the twentieth century."

In short, dramatic changes in science and in communication are forcing changes in science communication, and, in the process, in the kind of scientist who gets communicated. Today's visible scientists, while resembling their predecessors in many ways, are unique to contemporary media and their audiences. And if media continue to evolve as predicted, visible scientists will evolve with them.

What has happened to our good old-fashioned image scientist? He has become a relic, a holdover from a previous journalistic era, an image that has failed to change at the future-shock pace of technology. Even Einstein's image has been used lately to parody scientific genius. Like a bulldozer, electronic communications have leveled American images, substituting media celebrities for heroes. Politicians, movie stars, gurus, athletes — and scientists — are impaled under a magnifying glass and required to share not only their achievements but their families, their idiosyncrasies, their past sins, their favorite foods. The limelight is hot and harsh, more like a policeman's interrogation light than a shimmering stage light.

No longer prepared to idolize science-heroes, Americans and their media have created science-celebrities. Emphasis is on entertainment, the visual, the spontaneous, the thespian; education is merely a by-product. Even in newspapers, it has been estimated that after advertising (which alone is seventy-five percent of the average paper), sports, comics, society news, human interest stories, and stories about movie and television stars, one-sixteenth of the average newspaper is left "to cover and comment upon the world of governments and politics; wars and revolutions, defense and diplomacy; public health, welfare and justice; science, invention, religion, education, literature and the arts. . . ." Of this scant one-sixteenth, about two percent is science news — or about one-tenth of one percent of the total newspaper. On television, with the exception of scarce documentaries, hard science news must compete with the top stories of the day for a place in the half hour of time on the evening news. If science information doesn't make the "front page," it doesn't make it.

The would-be scientific hero becomes lost today not only in the celebrification process but also in the fragmentation of his profession. Each subspecialty of science has its own subheroes. A story circulates in Washington that William Shockley, one of three engineers from Bell Telephone Laboratories to win the Nobel Prize in physics for development of the transistor, encountered at the National Academy of Sciences a member he did not know. In response to Shockley's inquiry, the affronted scientist replied, "I'm known as the father of cortisone." "Well," Shockley returned, "*I'm* known as the father of the transistor." In politics as well, no one scientist or even a few can claim to represent a consensus of the large (about 1.7 million) and fractionated scientific profession.

Today's visible scientists are those who are adjusting to the modernization of science communication, to the changes in communication technology and in the scientific community. In the survival struggle of our rapidly evolving communications media, they are the scientists who are adapting. In fact science writers frequently cite them as the most influential force in contemporary science news.

Part of our government-by-crisis, visible scientists are catalysts in the process of converting problems into visible issues. As figureheads, they attract media (who follow public opinion) and politicians (who follow the media and public opinion), coalescing public concern and precipitating changes in national priorities.

Because they are scientists, the visible scientists also draw science into the political fray, prodding political attention toward scientific and technological issues — the most pressing issues of our time. According to one estimate, over half the bills before Congress now have a scientific or technological basis: health, energy, food, natural resources, environment, product safety, outer space, transportation, communications. Said the late Jacob Bronowski, "The world today is made, it is powered by science; and for any man to abdicate an interest in science is to walk with open eyes toward slavery."

Science has produced communications technologies that have democratized learning and experience; now the public demands the democratization of science. The public expects to participate, and indeed must participate, in determining the directions of science. Elitism is no longer tolerable — it is no longer safe. Consciously or unconsciously, the visible scientists are combatting elitism, exploring the new media, experimenting with ways to provide the kind of information the public now demands.

The scientific community is as uncomfortable about the democratization of science communication as the rest of us are about some of the other effects of technology. Like any modernization, the updating of science communication is slow, painful, often resisted. The new visible scientists are seen by their colleagues almost as a pollution in the scientific community — sometimes irritating, sometimes hazardous. The new scientists are breaking old rules of protocol in the scientific profession, questioning the old ethics, defying the old standards of conduct.

As a result, the visible scientists are an unusual group, having emerged from strong cross-pressures — pressures from the scientific community to concentrate on their scientific research, pressures from the public to provide input on critical issues, pressures from the press to conform to standards of newsworthiness. And, unlike politicians or movie stars, their very success may have the opposite effect on their research careers.

They deserve a closer look. Individually, contemporary visible scientists have been much discussed, idolized, cursed, applauded, and ridiculed. A few have been studied in detail. This is a book about the visible scientists as a group. It began as a doctoral dissertation in Stanford University's Department of Communication, a study of the interrelationships among the scientific community, the mass media, the visible scientists, and the public. After identifying forty-five visible scientists on the basis of two surveys, I selected eight of these scientists for special attention — Paul Ehrlich, Barry Commoner, Glenn Seaborg, Linus Pauling, B. F. Skinner, Margaret Mead, Carl Sagan, and William Shockley — and they became the subject of profiles illustrating the main points of the book. Some of the methodological background of the dissertation can be found in chapter notes at the back of the book; there is much more detail of course in the dissertation itself.

Climaxing the five years of research were over one hundred interviews in the offices and homes of visible scientists, their friends and colleagues, and the reporters who write about them. In conversations as dramatic and unpredictable as the visible scientists themselves, the rapidly changing visibility system was explored: What kinds of scientists now reach public attention?

What role do the media play in singling them out? What forces within science bring them to visibility? What do their more traditionally minded colleagues think of them? Do their scientific careers suffer? What influence do visible scientists in turn have on the media? Do they deliberately cultivate publicity? How serious are the many criticisms leveled against them?

The Tribal Scientist

Our time is a time for crossing barriers, for erasing old categories — for probing around.

— MARSHALL McLUHAN

"The battle to feed all of humanity is over. In the 1970s the world will undergo famines — hundreds of millions of people are going to starve to death in spite of any crash programs embarked upon now. . . . We can no longer afford merely to treat the symptoms of the cancer of population growth; the cancer itself must be cut out. Population control is the only answer."

With such predictions, pleas, and proposals, between 1968 and 1970 Paul Ehrlich, a Stanford University biologist, captured the imaginations of middle-class Americans. His book, *The Population Bomb*, sold over three million copies. He became known as the Ralph Nader of ecology, the angry young man of the environment movement, a leader in the international crusade for population control.

In 1970, at the climax of his visibility, he received over two dozen requests a day for personal appearances, although he charged $2,000 for lectures and was booked a year in advance. Reporters seeking two-hour interviews were granted ten minutes. *Ramparts*, *Playboy*, and *McCall's* were after him to submit articles, and he was featured in *Life*, *Look*, and the *Washington Post*. Television crews crowded his office, filming his opinions on Nixon, abortion, the pope. Johnny Carson featured him on the "Tonight" show for an unprecedented sixty minutes, although "the show isn't supposed to be serious"; the broadcast elicited a record five thousand letters to NBC, and Ehrlich was invited back within six weeks. On the run eighteen hours a day, eighty thousand miles a year, he left his research largely in the hands of two graduate students and postponed publishing the results. He was so pressed at one point he "fled the country" with his graduate students for a relaxing month of field work in Trinidad.

Paul Ehrlich is a visible scientist of the 1970s. A full professor at Stanford University, his credentials as a researcher in entomology (the study of insects) and population biology make him unquestionably a "scientist." He has published several textbooks and over one hundred technical papers, with titles like "The integumental anatomy of the silver-spotted skipper, *Epargyreus clarus* Cramer (Lepidoptera: Hesperiidae)" or "Problems of articalpine insect distribution as illustrated by the butterfly genus *Erebia* (Satyridae)." He "grew up chasing butterflies and dissecting frogs" and loves his research. He has been a member of the editorial board for journals like *Systematic Zoology*, a member of the executive council for the Lepidopterists' Society, a fellow of the California Academy of Sciences.

But he is also a best-selling author, a popular lecturer, a television personality, and a consummate salesman. The public has no idea he is an entomologist: they know about *The Population Bomb* and his vasectomy. (He underwent the sterilization operation about 1963; his one child, Lisa Marie, was born in 1955.)

Not the shivering scientist who withers in front of camera lights, Ehrlich is at ease before audiences, with reporters, on television, in print. Whether speaking to one reporter or one million television viewers, he combines an uncanny way with words, a flair for the dramatic, a dash of well-placed exaggeration, all communicated at breakneck speed. Urgency is the impression he leaves, softened by honest, casual warmth. (Regretfully, he is off to catch another plane; he would spend hours with you if he only had the time.) Young (in his forties), tall (six-feet-two), dark (and tan), he reminds people of Ralph Nader in appearance as well as approach. Reporters like to call him lanky, and rumpled. He has short hair and long sideburns — "I get away with a lot this way."

Much of his charisma comes from a writing and speaking style that is natural, quick, caustic, iconoclastic, hyperbolic, vivid, witty, and anecdotal. Examples:

Q: Then, you really think this population growth is that serious?
A: I wouldn't say serious; I would say fatal.

(National Wildlife, 1970)

Give your child an IUD to take to "Show and Tell."

(The Population Bomb, 1968)

Nothing distresses the victims of the Protestant ethic more than the thought of people leading a less hateful life than they lead.

(How To Be a Survivor, 1971)

Saying that the population explosion is a problem of underdeveloped countries is like telling a fellow passenger, "Your end of the boat is sinking."

(Reader's Digest, 1969)

In the case of apparently uneducable politicians, such as Governor Reagan, groups of scientists can band together to monitor, collate and analyze any absurdities they perpetrate.

(Speech to the American
Physical Society, 1972)

No one today can deny that the United States has the grossest national product in the world.

(Speech entitled "The
Environmental Crisis")

There are 3.6 billion of us. According to every estimate that can be made, that is somewhere between 3 and 7 times more than the planet can possibly maintain over a long period of time. . . .

(Speech entitled "The
Population Bomb")

People haven't wanted air quality standards to go down like Mr. Ford [President Gerald Ford] does, but of course people aren't owned by industry, Mr. Ford is.

(Interview for KPBS television,
San Diego, 1975)

Ehrlich "shoots from the hip," as an assistant put it, and qualifies, if need be, later. Associates around him act as moderating influences (with his blessing), and try to prevent anything too rash from finding its way into print.

Explaining his free-wheeling and far-reaching approach, Ehrlich describes himself as being in the "Professor von Stoopnagel syndrome." Sid Caesar used to do interviews as a Professor von Stoopnagel, Ehrlich recalls. In one episode, the professor is mountain climbing. The interviewer asks, "What do you do if you fall off a cliff?" The professor answers, "I put my arms out like this." (Ehrlich demonstrates, flapping his arms.) "But Professor, man can't fly," the interviewer objects. The professor replies, "Who knows, I might be the first!" Explains Ehrlich, "Ecologists are going like this [arms flapping], but what the hell else is there to do?"

Ehrlich feels that, with so much at stake, he must express his opinion, even at the risk of being wrong. "You do your best, express your opinions, and take your lumps." Democratically, he is equally willing to administer "lumps," to "not-too-bright politicians, . . . the mental defectives that run automobile companies," and other ecological foes. The world for Ehrlich, according to a friend, is divided into good guys and assholes (also known as dumdums or yoyos). Without batting an eye, Ehrlich took a swing at a large portion of today's scientist-leaders in a 1975 article he wrote for the *Bulletin of the Atomic Scientists*, "The Benefits of Saying YES!" The article was a tongue-in-cheek eulogy to unbridled technology, a take-off of an earlier article, "The Hidden Cost of Saying NO!" by physicist Freeman Dyson of the Institute for Advanced Studies at Princeton. At the end of the parody, Ehrlich placed a list of "Acknowledgements": "I would like to thank Thomas Andrews, Wilfred Beckerman, Hans Bethe, Wernher von Braun, Colin Clark, Buckminster Fuller, Philip Handler, John Maddox, Viscount Pirrie, Dixy Lee Ray, Edward Teller, and Alvin Weinberg, whose thinking on technology has been a constant inspiration to me in the writing of this paper — P. E."

Ehrlich arrived at his position as the Ralph Nader of population by choice and chance. Having developed an interest in ecology as an undergraduate at the University of Pennsylvania (an interest he traces to reading William Vogt's *Road to Survival* as a freshman in college), he emphasized population studies in his graduate work at the University of Kansas, considered at the time to have the best Department of Entomology, and in later research. He began including discussion of his population concern in courses he taught at Stanford University. Students mentioned him to parents, and he was soon invited to speak at alumni gatherings. From alumni, it was an easy step to other groups, including finally the prestigious Commonwealth Club in San Francisco, where his speech was covered by radio and television. "I had made a decision to try and have an influence in this area. All teachers want to influence people, and that was a rational, calculated decision. However, what was not rational and calculated was the result of having my mouth open at that particular point in time, when apparently people were ripe to listen. And all of a sudden I was very much in demand for Bay Area things."

Ehrlich's Bay Area media appearances attracted the attention of David Brower, executive director of the Sierra Club, California's largest conservation organization. Brower, together with Ian Ballantine of Ballantine Books, suggested that Ehrlich write a book. "We naively thought such a book might influence the 1968 presidential election" by pressuring candidates to take a stand on the population issue, Ehrlich recalls. Ballantine Books promised to get the book out in a matter of months, just in time for the election, so Ehrlich hastily produced *The Population Bomb.*

In the process of "schlepping" (promoting) *The Population Bomb* for Ballantine ("I thought, what the hell, it's an opportunity that not many biologists get, to go out and present our view"), his visibility naturally increased. And his life began to include a larger element of serendipity. Arthur Godfrey sent the book to Johnny Carson, who decided to take a chance. The startling success of that encounter "just sort of destroyed my life right there." It forced him into a position where it was much harder to fade back into obscurity. His research in this period would have gone "down the tubes," he says, if he had not had bright graduate students to take over his laboratory for a while.

The fad and flurry have receded since the peak in 1970, but Ehrlich still receives about one request a day for a personal appearance. He was tempted to quit public life altogether after the 1972 election, but has compromised on a steady, lower level of activity. Since he hates "ricocheting around the country . . . eating too much lousy food on airlines . . . getting drunk when you really don't want to get drunk," he consolidates speeches into two or three tours a year. He reserves days for research and teaching and does his popular writing on evenings and weekends.

The whole process of accumulating visibility took for Ehrlich only three years. In 1966, four years before his 1970 peak of activity, he was an unknown, although popular, lepidopterologist at Stanford University, a young-man-on-the-way-up in biology circles. Although he had begun at Stanford as assistant professor only six years previously, he had just been promoted to full professor. He had been publishing technical papers since he was sixteen years old, and was considered at the forefront of the new field of population biology. He had accrued appropriate honors and had been secretary of the Lepidopterists' Society. He seemed devoted to his work, and often took field trips to escape from the pressures of his job. When a colleague visited his office, he might take out a jar of butterflies that had been relaxed and pin them while chatting. His hobby was flying; he was a licensed pilot who averaged one hundred hours a year. (A sign in his office still says, "I'd rather be flying.") He was a popular teacher, partly because he interjected discussion of the "human side" of the population studies into his lectures; he had published two general articles on the subject in little-known magazines: *Chemistry*, a journal for high school teachers, and *Stanford Review*, an alumni magazine.

What happened? Why the sudden public visibility? The answer lies partly with the media. In an age of live television, telephone lectures, and mass paperbacks, Ehrlich is a natural.

He is a McLuhan man, a priest in the new global village tied together by communications technology, at home chatting on television, fielding reporters, fashioning magazine articles. In an era when it is increasingly difficult for most people to get access to media, Ehrlich had no trouble: they came to him. And he took advantage. "Any loudmouth, if, instead of talking to two hundred students at a time can talk to two hundred thousand over radio and TV, obviously is going to take the opportunity."

Paul Ehrlich's father was a shirt salesman, his mother a school teacher, and in Paul, friends say, you see the combination. His approach to the media is aggressive: "If you're going to win this game, you're going to go about it systematically, by organizing people, by influencing kiddies when they're very young, by belting them in every possible way, in every direction, with whatever techniques or whatever else tends to work. Of course, we are very much limited by numbers and by money. If we had a two-million-dollar-a-year budget, we'd have the problem well solved. We'd just spend it in the media. Just think what people in communications can do."

Debates, speeches, magazine articles, interviews, technical papers, and college and high school textbooks were all part of a conscious effort by Ehrlich's group to get the message about population control to as wide an audience as possible. *The Population Bomb* was revised in 1971, and a second Ballantine paperback added. The second book, *How To Be a Survivor*, was coauthored by a law student assistant, Richard Harriman, and directed to a young audience. In 1974, a third Ballantine paperback was added, coauthored by Ehrlich's wife, Anne, *End of Affluence*, a sober look at the realities of scarce natural and economic resources. Earlier in 1974, Ehrlich and a young political scientist, Dennis Pirages, produced *Ark II*, a political and social analysis of the causes of the present ecological crisis. In the meantime there have been high school and biology texts, including *Population, Resources, Environment: Issues in Human Ecology*, again with Anne Ehrlich, first published in 1970, revised in 1972 and 1976.

Paul Ehrlich had in essence been selected by the media from among the scientists who would be willing to play his public role. Sensing this, a cluster of colleagues who supported the population cause at Stanford worked out a division of labor: Paul was the one, they explained, to do the Johnny Carson show, or the Commonwealth Club — he had a "talent" for popularizing; collaborator and friend Richard Holm was good at organizing textbooks and digging out the tedious details that bored Ehrlich; John Thomas, a Catholic, spoke to church groups. Ehrlich's wife, Anne, frequently worked in Ehrlich's office, pulling his writing into shape ("She tells me it reads like gibberish"); in recent years she has become a full collaborator, although her name on the book jacket of *End of Affluence* is set, at Ballantine's insistence, in type half the size of her husband's. Ehrlich lavishly acknowledges his silent partners, although friends says he likes being the center of attention, discourages assertiveness around him, and dominates conversations. If he likes to hear himself talk, the public concurs. At one time everyone in the group at Stanford used to give popular lectures, but it was soon apparent, one explained, "they don't want anyone except Ehrlich."

Other scientists have undergone a similar screening process. As in natural selection, a special species of scientist, a small group within the scientific community, has evolved which is "fittest" for the media. And, as in natural selection, the new group of scientists is selected according to certain key traits, the traits the media now favor.

Just what it takes to make a science-celebrity is evident from a look at the "fittest" in the species — today's visible scientists. In spite of their remarkable diversity, the visible scientists share several important media-oriented characteristics, which stand out when they are pictured as a group. Most of the characteristics are naturally shared by other celebrities, but they are given a special twist by the fact that the newsmakers are scientists. Five characteristics are especially salient: the visible scientist is relevant, controversial, articulate, colorful, and reputable as a scientist.

The Visible Scientist Has a "Hot Topic"

The magic word these days is "relevance." Students want their school work to be relevant to their problems; government wants scientific research to be relevant to national needs; readers want newspaper information to be relevant to their daily lives. On this assumption, at least, the press features local news, sports news, a women's page. And in the area of science, editors emphasize practical applications and policy implications rather than abstract research itself. They look for pollution solutions, not particle physics. Much of what used to be "science reporting" has become "science policy reporting."

Today's visible scientists are speaking out in areas of general concern, talking not about their research but about its ramifications. They are at the forefront of issues that count and that catch attention. When René Dubos, the Rockefeller University microbiologist and environmentalist, saw a list of today's widely known scientists, he said, "This list reflects what problems are in the minds of people today." Said Margaret Mead, "What you've got here is the environment, the management of human behavior, and the bomb. That's about it . . . and it's all issues."

To a certain extent, this means the visible scientists are a product of media fads. Topics move in and out of vogue, and with them the scientists associated with them. In the late 1950s, the hot scientific topic was fallout; in the early 1960s, disarmament; in the later 1960s, the ABM and the SST, population, and environment; in the early 1970s, energy.

Timing is important not only in getting on the bandwagon but also in getting off. Deliberately or instinctively, scientists with lasting visibility and influence have moved from one issue to another with shifts in public interest. They are on the next peak before the previous one becomes a trough. Paul Ehrlich has recently taken up the issue of nuclear power safety. Linus Pauling, who has been in the limelight on and off for over twenty years, was known to the general public first for problems in the McCarthy era stemming from accusations that he was a Communist, then for protest against nuclear testing and fallout, and most recently for his theories on vitamin C. Probably the most dramatic example of staying power, Margaret Mead, a public figure for almost fifty years, has addressed herself to topics ranging from sex differences to the generation gap and, recently, to environment.

Even straight science topics move in and out of vogue. Physics used to be the glamor field — atomic bombs, nuclear fission. Now there is much more emphasis on biological sciences — cancer, genetic engineering, immunology. To some science writers and editors, Walter Sullivan, *New York Times* science editor and "dean" of American science writing today, is an anachronism because he specializes in stories from the realm of physics. Indicating the intensity of fads, the managing editor of a major daily said, "If Sullivan were to offer to work for me today, I wouldn't hire him. He's a bore."

New fields become favored by the press for a number of reasons. Exobiology (study of extraterrestrial life) and oceanography have been considered by reporters to be at the "cutting edge" recently. In contrast to physics, for instance, theory in these fields is being changed and expanded at a rapid rate, and broad, new discoveries are being made that capture the imagination. The glamour of these fields, reporters believe, has increased the visibility of scientists within them — for instance, Carl Sagan, Cornell University astronomer, and Roger Revelle, oceanographer and Harvard University professor of population policy.

With straight science, as with science issues, however, one of the key ingredients is relevance. The "humanness" of her topic plays a role, for example, in the visibility of British ethologist Jane Goodall (Baroness Jane van Lawick-Goodall), one of the very few scientists to have achieved visibility in the last ten years for scientific work rather than policy discussion. Called "the world's foremost authority on chimpanzees," she is also a popular lecturer and author, narrator of several wildlife documentaries for American television, and a former visiting professor in psychiatry and human biology at Stanford University.

There are of course a number of reasons for Goodall's popularity. For one thing, her story has romance: she has lived for over twelve years in the remote Gombe Stream Reserve in western Tanzania, East Africa. When she first set up camp on the game reserve, she was twenty-six; for five months her mother stayed with her, because the authorities would not allow a young English girl to wander around alone. With remarkable patience, she observed the chimpanzees all day long, month after month. In 1962, the National Geographic Society sent a Dutch photographer, Baron Hugo van Lawick, to film her work. They fell in love during his assignment and were married in 1964, returning to work together at the camp after a honeymoon in Europe. (They were divorced recently but continue to collaborate professionally.) Hugo, Junior, nicknamed "Grub," was born in 1967 and has grown up with his mother in Tanzania. Goodall, who left school at eighteen, got started on her striking career when she met the famous anthropologist Louis S. B. Leakey while she was visiting a friend in Kenya. Leakey hired her as a secretary at the Coryndon (now National) Museum in Nairobi, where he was curator, and encouraged her interest in animals and Africa, suggesting she study chimpanzees. He secured her first funding for the Gombe project and arranged for her to write up her results as a Ph.D. dissertation for Cambridge University, making her one of a handful of students in the history of Cambridge allowed to work for a doctorate without first taking a B.A. degree. Leakey also, according to Goodall, was the one who arranged for Hugo to be assigned to photograph the chimpanzees. Goodall went on to make major contributions to her field, brushing away musty theories on the basis of her observations of chimpanzees in their natural habitat.

There is, then, the "corny business," as she calls it, "that I happen to be a woman and I did something offbeat — that interests people of both sexes and it particularly interests young girls. I've had hundreds of letters from young girls, saying, 'How can I be like you?'" The *Ladies' Home Journal* sent actress Candice Bergen to Tanzania to interview the actress's "hero." Other segments of the public are captivated, startled, and a little uncomfortable, envisioning a young, slender, "ethereal" (in a *Saturday Review* description), blond girl, far from civilization, mingling with the primitive chimps. There is a sense of the forbidden, and the ultimate in female liberation: a woman alone, in the wild, tending a son, coping with malaria and cobras, building a successful scientific career. A man in her position would probably be ignored.

Goodall also attributes her visibility, about which she was "fairly surprised at the beginning," to a kind of social relevance. People are fascinated, she has found, by the similarities between chimps and man. "The fact that something that isn't human can behave in a way that is so like man surprises people," Goodall says, "and they look at themselves in a new way — or they look at the chimps a new way — one or the other; it doesn't much matter which."

Her lecture audiences are often composed of parents, who chuckle at the similarities between baby chimps and human children. They ask her whether she picks up ideas on child rearing from chimp mothers — she does, and plans a book for human mothers based on her animal observations. Young people are also attracted, Goodall believes, by "primitivism in the decent sense of the word . . . a getting back to nature, away from pollution." Concerned about man's "headlong race to destruction," they look in her work for answers to why human beings behave as they do. Reporters and television writers emphasize the ties to human life, although she personally is interested in "the chimp as chimp." She is also the author or coauthor of several popular books, including *My Friends the Wild Chimpanzees; Innocent Killers;* and *In the Shadow of Man,* as well as technical articles.

The ultimate example of a "hot topic" is, of course, sex. William Masters and Virginia Johnson, the St. Louis sex researchers, wrote *Human Sexual Response* for professionals, not for the public, and they sold it as an ordinary textbook. "It was written in eye-wrenching gothic style," according to writer Patrick McGrady, Jr., "expressly to refute charges of pornographic intent and preoccupation." In fact, the lower-than-expected sales of the second book in the series, *Human Sexual Inadequacy* (1970), probably reflects the lay public's disappointment with the dryness of the first. Both were nonetheless best-sellers, and Masters and Johnson are as widely known as any researchers today.

The Visible Scientist Is Controversial

For better or worse, it is a habit in the news media to emphasize drama and conflict, and to highlight the controversial aspects of news stories. If hecklers disrupt a speech, the hecklers, not the speaker, get the headlines. If two selectmen lock horns at a town meeting, their argument gets the coverage. In the story on a presidential candidate's daily activities, his most startling statements will take the lead. In science news also, controversy is an increasingly frequent element.

It is no coincidence, then, that today's newsmakers, including scientists, are those who have a controversial component. A safe stand on a safe topic simply does not make news. As reporters put it, science per se as a topic is "out," but controversy is always "in."

The visible scientists are those who are willing to take unqualified, dramatic stands on issues. Paul Ehrlich, for example, took the position that population was the fundamental environmental problem, underlying resource scarcities, health problems, and pollution. Thus he laid himself open to challenges from the church, social scientists, blacks, the Third World. Economists brought up the role of economic forces in encouraging pollution. Sociologists, assuming that social forces would correct the biological trends toward overpopulation, challenged Ehrlich's belief that they would not. Ehrlich's extremes elicited equal and opposite extremes from the opponents of "ecological crisismongers." A San Francisco columnist, Charles McCabe, calls him "a latter-day Malthus" and "worse than Hitler." Said *National Wildlife* editor John Strohm, after publishing an interview with Ehrlich, "No single article ever published by *National Wildlife* has provoked more letters, more requests for reprints, more pro and con reactions."

The visible scientists are mavericks, frequently at odds with fellow professionals, the academic establishment, the political system, religious tenets, and prevailing public sentiment. *Time* magazine calls B. F. Skinner "the most controversial contemporary figure in the science of human behavior, adored as a messiah and abhorred as a menace." Linus Pauling took on McCarthyism and the military establishment in the 1950s and 1960s, the medical establishment in the 1970s. Arthur Jensen, William Shockley, and Richard Herrnstein run afoul of religious principles, the democratic ideal, and trusted rules in education; their confrontation has resembled a battle between heretics and the state more than it has a debate among scientists.

The visible scientists are gadflies, then, both within their profession and in society at large. In the terminology of political scientists, they are among the few "gladiators" in the political fray. They are usually not political extremists — scientists on the extreme left or right lose credibility with the press. Politically, most visible scientists are described by acquaintances as "liberal," not moderate, but not extreme. And when they take stands on particular issues, they tend to occupy the same range, away from both the middle and the far fringes. Their position is thus usually interesting, debatable, but not frightening or absurd.

The visible scientists are controversy prone, the kind of people who look for new approaches and, finding them, advocate change. The same tendency is apparent in their scientific work: they are revolutionaries, questioning established theory, proposing new concepts. Paul Ehrlich has a reputation for being innovative in his field, "always on the cutting edge"; a 1962 article coauthored by Ehrlich in *Science* is credited with bringing together "Population Biology" as an area of study, giving the field coherence and unity. A later article coined the term "coevolution," a concept gradually adopted by biologists. Papers coauthored by Ehrlich have questioned and scrutinized such trusted biological concepts as the species, gene flow, and the natural selection process. Explaining their approach in prefatory remarks to the 1962 paper, which questions most of the basic tenets of earlier population biology, Ehrlich and Holm write, "In the discussion which follows it may seem that we have restricted ourselves largely to destructive criticism, demonstrating the disadvantages of established procedures and modes of thought. But these must be pointed out before it is possible to develop improvements. . . . In science it frequently is necessary to criticize existing theoretical structures to clear the way for new ideas." In the process, Ehrlich has attracted opponents and bitter dispute, and made "life-long enemies" of many scientists, he believes.

Scientific and public controversies abound among visible scientists. Ehrlich and Washington University biologist Barry Commoner had a kind of summit dispute of environmentalists in the early 1970s. The Olympian struggle began as a difference of opinion over the relative importance of population control in solving the environmental problem: Ehrlich maintained that population control is a prerequisite to curbing environmental deterioration; Commoner argued that poorly managed technology is the villain in environmental deterioration. The argument between the two environment spokesmen quickly left the confines of intellectual debate, however, and broadened to include personal digs, condescension, and chicanery.

Trouble began in 1970. No one remembers when the two scientists first met — probably in 1969, at a meeting in San Francisco, or at an ecology seminar at Stanford. Early in 1970, physicist John Holdren, a close associate of Ehrlich, attended a private gathering of environmental experts convened by Bernard Berelson, president of the Population Council, in New York City and took back to California the message that Commoner had loosed a tirade against Ehrlich at the meeting. The first and only public confrontation between Commoner and Ehrlich occurred in December 1970 at a symposium organized by Commoner at the American Association for the Advancement of Science (AAAS) meeting in Chicago, in which Ehrlich and population spokesman Garrett Hardin squared off against Commoner and demographer Ansley Coale on the role of population growth in environmental deterioration. Reporters found the debate acrimonious and confusing — "they were screaming at each other," recalls one writer.

By 1971 both authors were criticizing each other by name in their writing. Said Ehrlich and coauthors in the preface to a collection of readings, *Man and the Ecosphere*, "Some commentators have erroneously implied that our environmental problems can be traced primarily to misuses of technology since World War II," adding in parentheses, "(see, for example, Barry Commoner, *The Humanist*, November–December 1970, p. 10)." One finds, also, that when Ehrlich and his wife revised their textbook, *Population, Resources, Environment*, the 1972 version was rougher on Commoner than had been the original in 1970. Commoner pressed the view, in speeches and writings, that Ehrlich, Garrett Hardin, and others were proposing to coerce the public into practicing population control. In *The Closing Circle*, Commoner writes, ". . . some population-minded ecologists hold that 'we must have population control at home, hopefully through a system of incentives and penalties, but by compulsion if voluntary methods fail.'" Chapter Notes attribute the quotation to Ehrlich's 1968 *The Population Bomb*. "The outcome," Commoner continues, "would be to constrain, by compulsion, the public choice between the two paths toward social progress. More simply stated, this is political repression." Although Commoner says he tried to confine his remarks to the issues, Ehrlich felt he was personally attacked, and by 1972 both sides were throwing personal barbs.

Fanning the fire were incidents like Ehrlich's January 1972 speech to the American Physical Society, in which he mentioned Commoner in the same breath with the unpopular Shockley as scientists typical of a "dangerous trend" toward one-sided public appeals. Ehrlich argued that although the public had liked Barry Commoner's new book, *The Closing Circle*, professional ecologists were horrified by Commoner's thesis; Ehrlich added: "I predict that politically active scientists will more and more appeal to the public for support when they receive little or none within their professions. In my view, this is a most dangerous trend, typified today by Commoner's unidimensional treatment of the complexities of the environmental crisis and by William Shockley's racial crusade, both of which have negligible support among scientists competent in these fields." This statement, later quoted in a press release, was considered highly inflammatory by the Commoner camp. Ehrlich later told *Science*

magazine that the comment was "perhaps open to misinterpretation." Ehrlich's assistants were mortified that they hadn't caught
the potential problem before the hastily-prepared speech had
been delivered.

On the other side, Ehrlich aides say that later in 1972, at the
United Nations Conference on the Human Environment in
Stockholm, Commoner masterminded a plan to subject Ehrlich
to embarrassing harassment. During one of the panel sessions at
the meeting, Third World representatives overwhelmed Ehrlich
with accusations about his views on the Third World, accusations based largely on his statements in the first edition of *The
Population Bomb.* Paul Growald, Ehrlich's executive assistant,
says he was told by several people that during the episode Commoner sat on a balcony and prepared nasty questions for Third
World participants on the floor to fire at Ehrlich; Commoner
denies the story, but agrees that he was in the balcony and that
he spoke with many Third World delegates at the meeting.
Growald also says that Commoner refused an invitation to dinner with Ehrlich.

As the controversy wore on, arguments broadened to include
spurious issues: Ehrlich's motives, Commoner's qualifications as
an ecologist. Each side grew more extreme. *Science* magazine
assessed the situation in 1972 this way:

> The bulk of opinion among physical and social scientists seems to
> be that both parties are carrying their views to the extreme, Ehrlich
> being an "alarmist" and Commoner strangely obstinate in exonerating
> population growth. Demographer Ansley Coale, for example, is not
> crazy about either approach and believes that ideological commit-
> ments on both sides are obscuring the scientific questions. Com-
> moner "is sort of mystical about the balance of nature but somehow
> absolves the role of population growth," he says, while Ehrlich, "a
> real missionary on ZPG (zero population growth)," is regarded by
> many demographers as being to population what the Women's Chris-
> tian Temperance Union is to alcohol.

The two sides could not even agree on how to argue: Ehrlich
made several overtures, privately and publicly, to "bury the
hatchet," on the grounds that it was counterproductive to air

the whole thing before the public, which would only be confused. Commoner, holding to his belief in the efficacy of public
debate, contended that the disagreement had a political element
which only the public, not an elitist tribunal, should judge.
Science concluded: "Ehrlich has been attempting to put the
debate on a purely scientific basis, whereas Commoner considers
politics to be very much part of the equation. When you're
playing bridge and your opponent's playing poker, it's hard to
agree on the rules."

The controversy climaxed in an incident in which the editors
of *Environment* magazine, with which Commoner is closely associated, apparently pirated an article by Ehrlich and published
it without permission in *Environment*. Early in 1972, Ehrlich
and Holdren circulated a critique of Commoner's new book,
The Closing Circle. The paper was sent to a number of scientists
and journalists, including the Commoner group in St. Louis.
The Bulletin of the Atomic Scientists (also entitled at that time
Science and Public Affairs) arranged to publish the article and
scheduled it for the April 1972 issue. The *Bulletin* then invited
Commoner to write a rebuttal and suggested he publish it in
the May issue. At Commoner's request, however, the publication of the Ehrlich article was postponed until May, so that the
critique and rebuttal could appear simultaneously. It was quite
a shock to *Bulletin* editors and to Ehrlich, therefore, to find
that the April issue of *Environment* carried the Ehrlich article, based on the earlier unrevised and unedited preprint, and
Commoner's rebuttal. The articles in *Environment* in effect
"scooped" the *Bulletin* by a month.

A rapid exchange of letters followed, copies of which Ehrlich
later gave to an environment reporter. They show that there
were two points of view on the "scoop," but they suggest that
Ehrlich and Holdren had the stronger position. Letters shortly
after the incident from Richard S. Lewis, editor of the *Bulletin*,
and Harvard biologist John Edsall, a member of *Environment*'s
Science Advisory Board, chastised Commoner and *Environment*
for the move. The *Environment* staff responded by arguing that
in many important cases it is the function of the journalist to
bring information to public attention, whether or not the sources

of the information would want it published. Further, they stated, the Ehrlich–Holdren article was widely distributed, used as the basis of newspaper columns, and discussed in undergraduate courses; it had, according to *Environment* attorneys, no valid copyright.

Ehrlich and Holdren appear to have the last word in a second letter to the Science Advisory Board describing a conversation with Kevin Shea, scientific director of *Environment*. "Owing to *Environment*'s earlier performance in this matter," Ehrlich and Holdren write, "we took the unusual precaution of having Ehrlich's phone conversation with Mr. Shea monitored. It was heard by one of our colleagues and a secretary, detailed notes of the conversation were made, dated, and signed." On the basis of these notes, Ehrlich and Holdren quote Shea as admitting that he knew their article was scheduled to appear in the *Bulletin*, and that he, Commoner, and Novick knew *Environment* did not have permission to publish it. "As to the mechanics of doing it," says Shea, "yes, it was unethical."

Over lunch a year later, Commoner said the episode was "probably childish," recalling that Novick had wanted to do it, and that he had gone along. In keeping with his philosophy of public information, however, Commoner defends his role in the broader controversy, calling it in 1974 "part of a serious debate on a major policy," with important implications for developing countries.

A more common reaction to the dispute was that it was an appalling waste of time — a point on which Ehrlich agrees. Margaret Mead summed up the general consensus: "They both of them bore me to death. They are both right — if there weren't any people there, the technology wouldn't matter. One wouldn't be here without the other, and it's an infernal waste of time, the whole nonsense."

The Visible Scientist Is Articulate

"It's not so much what you say as how you say it," the saying goes; in science reporting, and most news reporting, it is both.

The press demands that the scientist offer some substance, and on a newsworthy, preferably controversial, topic, but his success will also depend on how he presents it.

While most scientists seem lost in the esoteric, the visible scientists have mastered the art of the exoteric. "They just have a knack," says science writer David Hendin, "that every line that comes out of their mouths is a good quote." When talking about science, they cut through the jargon and explain the essence of the matter clearly and vividly. They pepper their descriptions with well-placed analogies and metaphors, circumventing technical terms and leaving a lasting image. They see the broad picture and are able to put their own specialty into perspective, to explain where it fits into the whole scientific endeavor. They forego the details of methodology so important to scientific discourse, but dull, distant, and devoid of meaning to the layman. Says Ehrlich, "All scientific papers . . . are generally fakes. In other words, they're presented in a way which isn't the way the work is actually done — you know, 'Introduction,' 'Materials,' and 'Methods,' all that sort of thing. Well, of course, popular audiences switch to another channel by the time you get to 'Materials' and 'Methods.' "

Visible scientists' way with words is a result of talent and training, with emphasis on the talent. While he calls writing "the hardest, most miserable work I know," Ehrlich put together *The Population Bomb* from lecture and article notes in just three weeks. And he shows a special interest in writing: "for the hell of it," he might try writing a novel sometime. He has written one fiction story, "Eco-catastrophe!" picturing the consequences of present population policy, and there are fictionalized scenarios in *The Population Bomb*. Before his visibility, Ehrlich contributed an article to *Flying* magazine, a personal account of a near miss early in his private flying career, for the magazine's reader-contributed section, "I Learned about Flying from That." And words seem to come as easily on the podium as in print. A Stanford demographer, Dudley Kirk, who had the misfortune to debate Ehrlich in the early days of the population movement, admits he was beaten hands down; even Kirk's wife, he says, agreed that he had made a poor showing.

Good writing comes more naturally to some visible scientists than others. Margaret Mead says she spins out three thousand words on a good morning; B. F. Skinner estimates he manages one hundred printable words on a usual morning. Those who find writing easy are enthusiastic: "I love to write," Mead says. Those who find writing harder are less exuberant. Says Barry Commoner, "Anyone who tells you it's easy to write is a liar." From Ehrlich, "I'd rather do almost anything else — it's either boring or hard as hell." He devotes "a tremendous amount of time," an assistant says, to working out metaphors so the audience can visualize a point. He gets many of his ideas from discussions with students; in animated bull sessions in the coffee room, he comes up with analogies, which he later incorporates into his writing. Ehrlich feels that most people could learn to write if they tried. "It's trained out of a lot of scientists."

The Visible Scientist Has a Colorful Image

In the March 1973 issue of the then *Saturday Review of the Sciences*, writer William K. Stuckey took on the task of writing a profile of John Bardeen, who had just won a second Nobel Prize in physics. Bardeen is a brilliant scientist — no one before had won the Nobel Prize for physics twice — but he is unprepossessing. He has none of the personality quirks of William Shockley, the quotable speaking style of Ehrlich, the dramatic history of James Watson. Stuckey finally wrote his article about just that fact: Bardeen is brilliant but unimposing. Stuckey wrote:

Against such a background of impressive figures [as appeared at Stockholm Nobel Prize festivities], John Bardeen might be easily overlooked. He is a squarish man of medium height, with flat eyes of an indeterminate color — probably blue or brown — covered by outdated plastic-rimmed glasses. At sixty-five, Bardeen is almost bald, with a ring of dull, straight hair about the back of his head. His voice, which he rarely uses, maintains a relentless monotone, changing pitch only when it cracks. There is nothing sharp or outstanding about him.

In spite of Stuckey's tack, Bardeen was a difficult subject for an article. Bardeen was uncomfortable and embarrassed about the article, in which he felt there were exaggerations and inaccuracies. From Bardeen's objections, it would seem that Stuckey's portrait showed Bardeen more colorless than he actually is: even his colorlessness is unextreme.

Bardeen is very likable, a sympathetic figure, as well as a "scientific giant," but he will never be a visible scientist. (And happily so; he finds the publicity hectic and distracting.) In these days of crowded streets and crowded newspapers, the scientist who becomes visible to the general public will be the one with a colorful image, which sticks in the reader's mind. A quiet scientist is no more likely to be featured than a quiet politician.

To make an impression on the public, a politician and his public relations staff build an image. On a less grandiose scale, "the making of the visible scientist" takes place the same way. Usually the scientist simply possesses certain striking characteristics which the press gradually molds into an image. Paul Ehrlich announces unabashedly that he has had a vasectomy. Environmentalist René Dubos has a grandfatherly charm and an expressive French accent (he emigrated in 1924) which is reminiscent of Maurice Chevalier. William Shockley tape records even casual conversations and phone calls. Linus Pauling is a political "radical" and wears a navy blue beret indoors and out.

The characteristics tumble together into a complete image. Out of four recent major profiles of Margaret Mead, at least three mention her cluttered attic office in a musty corner of the American Museum of Natural History, her dynamic energy and hectic schedule, her virulent critics, her motherly or grandmotherly role in the world, and her immense popularity as the leading public figure in anthropology.

To a certain extent, the flair of visible scientists is unaffected. But, consciously or unconsciously, they usually encourage the image. As they find certain things interest journalists, they mention them more frequently. Nobel Prize winners, writer Mitchell Wilson found, "get into an act," they are so used to telling the same story; one Nobelist practically has a number of skits, little routines he tells each time, even with the same inflection.

Visible scientists are also strong, assertive personalities. Egotism abounds. They dominate conversations, thrive as the center of attention, until even devoted followers and close friends admit they are "abrasive" — not the best characteristics in a spouse or roommate, perhaps, but the makings of visibility.

Journalism's *modus operandi* also encourages image-making. Linus Pauling recalls that when he won the Langmuir Prize early in his career, all the reporters asked what his hobbies were. He got tired of telling them he didn't have any, and, having always been interested in English language, started saying, "I collect dictionaries and encyclopedias."

A former biochemist who could give lessons in image-making is Isaac Asimov, until 1958 a full-time faculty member at the Boston University School of Medicine and today a widely known and respected science fiction and popular science writer. In his comments to newsmen and in his own writing, he helps his image along by encouraging people to think of him as a born writer, an egotist, and a ladies' man. On his writing abilities, he told *Publishers' Weekly* that he "types at professional speed. Ninety words a minute."

"Great, but do you think at ninety words a minute?"

"Yes, I do. The two work together neatly."

Articles always mention, too, that he works from nine to five, seven days a week, without a break ("No, I'm lying. Sometimes I goof off on part of Sunday"), and give the latest count on the number of books he has finished (over 160).

To cultivate a reputation for egotism, Asimov simply tells it like it is. He enjoys science fiction, he says, because he is part of a small, closely knit world in which he has "the status virtually of a demigod." In the area of popularizing science, Asimov also sees himself as a leader. Describing scientists famous for their popular lectures to the general public, including Humphrey Davy, Michael Faraday, Sir Arthur Stanley Eddington, he cited two recent examples, Fred Hoyle and himself. "I like to think

I am right up there in the mainstream of this facet of history. I imagine this will be, maybe it already is, taken for granted." Asimov's assessment of his position both in the science fiction and science popularizing fields is not inaccurate; it fits with the opinions of others. But it is, quite deliberately, immodest: "I assume," he said in an interview, "that my public image is that of a self-esteemed person."

Asimov adopted the ladies' man aspect of his image in the early 1970s, while he was between first and second marriages. Although his flirtatiousness went discreetly unrecorded by most journalists, Asimov hardly discouraged them from telling all. Explaining his decision to be interviewed for this book, he said, "I am strongly heterosexual and whereas I object very much to wasting my time having a man interview me, I have no objection whatsoever to having a young lady interview me." He is the author, under the pseudonym "Dr. A.," of *The Sensuous Dirty Old Man*, prompting a squib in *Sandworm* about his antics at a science fiction convention: "Dirty Old Men Still Need Lots of Love and Booze."

For Asimov, however, wine and women do not necessarily go together: he is allergic to liquor. A long-time friend and fellow science fiction writer, L. Sprague de Camp, says that he and his wife introduced Asimov to drink, with curious results: "About 1941, in our apartment in New York, we gave him his first drink. Asimov would have been 21. It was just a teensy drink, but he turned a funny purple color, with spots, and complained of feeling strange. After he left us, he rode the subway back and forth until he felt normal enough to go home. It wasn't inebriety but some allergy that got him. So, wise man, he swore off the sauce and has stayed off it ever since."

One aspect of his image Asimov could do without is the idea that he is a venerable sage, an aged if not deceased figure of history. In 1966, when he was forty-six, this part of his image inspired Asimov to write a poem entitled "The Prime of Life":

It was, in truth, an eager youth,
Who halted me one day.

He gazed in bliss at me, and this
*   Is what he had to say:*

"Why, mazel tov, it's Asimov,
*   A blessing on your head!*
For many a year, I lived in fear
*   That you were long since dead.*
*   . . .*
Since time began, you wondrous man,
*   My ancestors did love*
That s.f. dean and writing machine
*   That aged Asimov."*

I'd had my fill. I said, "Be still!
*   I've kept my old-time spark.*
My step is light, my eye is bright,
*   My hair is thick and dark."*

His smile, in brief, spelled disbelief,
*   So this is what I did;*
I scowled, you know, and with one blow,
*   I killed that rotten kid.*

The Visible Scientist Has Established a Credible Reputation

The *New York Times* science department receives enough mail each day to make a stack between one and three feet high, and the deluge at most major newspapers and television networks is similar. Science news items come not only through the mail, but also from phone calls and other informal sources. Somehow the reporter must cull from the masses of material a very few stories to write. And he must make his decision in a very short time.

Confronted with a potential story, reporters and editors must weigh a number of factors, including the credibility of the scientist in the story. An unknown scientist is risky. He may simply be a publicity-seeker, and his claims may be fraudulent. Every established science reporter can remember a case in which he covered a story that turned out to be a hoax. Walter Sullivan recalls a doctor at a fund-raising dinner who had a patient rolled into the dining hall, announced that the man's severed spine had been reconnected, and, in a scene parodying the Bible, commanded him to get up and walk. *Times* reporters checked the story, and were warned by their sources that it was medically impossible to reconnect spinal nerves after they had been severed. The *Times* treated the story carefully, but should have been even more careful, Sullivan says. It turned out the spine had been only partially severed, in which case it can often mend itself. Think of all the paraplegics, Sullivan laments, whose hopes may have been raised by the publicity-seeker's story.

Having been burned, science reporters are usually cautious. If they are not familiar with a scientist, they check with other scientists, about both the would-be newsmaker's personal reputation and his story. If the scientist has a good reputation in his scientific specialty, the process can be fairly quick; one or two sources will readily vouch for the newcomer's authenticity. If the scientist is an unknown among scientists, it will be harder to check up on him. And the reporter will have less time and inclination to cover the story.

The scientist with a reputation in his field, then, has the advantage. For this reason it is not surprising that most visible scientists were visible within their own fields before they became known to the public. Paul Ehrlich, although an unknown to the public when he first spoke out on population, had already contributed several new ideas to population biology.

The ultimate in a personal, checkable credential is the Nobel Prize. In fact, a Nobel laureate's views are not only welcomed but pursued. He has instant credibility with press and public. Of thirty-nine visible scientists studied in 1972, seven were Nobel Prize winners, and over forty percent were members of the prestigious National Academy of Sciences (NAS) or National Academy of Engineering (NAE). In an unpublished manuscript, Nobel laureate Joshua Lederberg quotes physicist C. J. Davisson as remarking in 1937 that a Nobel laureate is transformed "overnight from an exceedingly private citizen to something in the nature of a semi-public institution." Lederberg himself adds in the same manuscript, "The Nobel club has . . . achieved a distinctive notoriety, if not prestige, which is sustained mainly by the public press. It can be rather irritating to have one's individuality submerged under the inevitable attribution of the Prize at any introduction or reference in the press. And the club is constantly called upon to give its collective support to public issues that may have only the remotest connection with the scientific prestige of its members. I hope that thoughtful readers will not pass judgment too harshly on the idiosyncrasies of public behavior of many Nobel laureates if they first give some thought to the artificial pressures that have been focussed on them."

If a scientist does not have a wide personal reputation, he can get attention from the media by being associated with a widely known institution. The chances are, reporters figure, if he has been hired by Bell Laboratories, or Harvard University, he is reputable and safe. It can also be effective to speak from a reputable forum, such as a seminar or press conference at a meeting of the American Association for the Advancement of Science. Many stories about lesser-known scientists thus begin: "A Stanford University biologist announced today . . ." or "An American Physical Society committee expressed disapproval today of the administration plan . . ." Of the same thirty-nine visible scientists in the 1972 study, it turned out that four were affiliated with Harvard University, four with Stanford University, and four with the University of California — almost one-third of the group were from these three schools alone. None were from institutions whose names would not be familiar to virtually all science reporters and a large segment of the general public. Most also received degrees from widely known schools, and were located at major institutions at the time they first became visible.

One result of the media's demand for a credible reputation is that visible scientists are seldom young. The average age of the thirty-nine scientists in 1972 was fifty-nine; one, Carl Sagan, was under forty. Partly because the media have a credibility criterion, partly because young scientists tend to concentrate on research, and partly because lasting visibility takes time, the visible scientists are the ones with perseverance, and the ability to take setbacks and criticism without being discouraged.

That the visible scientists are also top scientists is not entirely a matter of media needs, of course. The characteristics that would lead to public visibility are often the same characteristics that make for success within a scientist's field. In fact, they are the same factors that make for success in any endeavor — ambition, energy, inquisitiveness, creativity, facility at explanations, organizational ability, aggressiveness, intelligence. "Society has a way of throwing up these innovators," Howard Simons remarked. It is the people with these characteristics who reach influential positions, not "the silent people — the people who need lots of sleep."

But the five characteristics of visible scientists discussed in this chapter show that the media are exerting a very real influence. The media have certain needs, and they are finding scientists to fill them. In the process, the media are of course shaping the kinds of scientists who will reach public attention. It is these scientists whose message gets to the public, across the complex channels of modern communication.

Guerrilla Science

How many of these people would you trust on a government committee?

— JAMES D. WATSON, Harvard University
biologist, commenting on a list of
contemporary visible scientists

In the late 1940s when high energy physicist Sidney Drell was a graduate student, he was taught to devote himself to his research, to spurn politics, and "not to hang around with the big shots in Washington." Today Drell, deputy director of the Stanford Linear Accelerator Center, commutes regularly to Washington to advise officials at the highest levels of government, and his students are embroiled in radical political causes.

In one generation, the role of scientists in politics has been transformed, not once but twice. In the 1940s most scientists stuck to their laboratories and avoided the taint of politics. In the 1950s it became fashionable to make occasional trips to Washington and give behind-the-scenes advice to government officials. In the late 1960s the behind-the-scenes "inside" advisory system lost its enchantment and effectiveness, giving way to a rash of alternative, "outside" activities in Congress, courts, and the press.

While "outside" activities were a traditional part of the American political process, they were anything but traditional for the scientific community. The scientific community had built up an established, acceptable way for scientists to discharge political responsibility: the inside, Washington advisory system. The increasing numbers of outsider scientists were guerrilla forces, fighting science's battles in unsanctioned ways, defying old rules, definitely not to be trusted on a government committee. Particularly disconcerting were the outsider scientists whom the general public found out about, the visible scientists.

Today's visible scientists are, then, an outgrowth of the rise and fall of the insider politics of the 1950s and 1960s. Coming at a time when the uneasy relationship between politics and science was in particular disarray, the visible scientists represented part of a new attempt by society and science to come to terms with each other.

Society has always been rather uncomfortable about science, maintaining an apprehensive ambivalence in which it was enthusiastic about the results of science but distrustful of its esoteric mysteries. And as scientists have become more specialized, they have become more mysterious. Increasingly science has seemed to take place behind closed doors, in the invisible worlds of the microscope and mathematical theory. Also, scientific values and findings have conflicted with religious and cultural tenets. Kenneth Boulding describes scientists as a small subculture in society, about the same size as the clergy; people view science as an "alien force," just as they would a minority, alien religion.

Societies have coped in different ways, each reaching its own uneasy equilibrium between the major culture and the alien scientific culture within it. Before World War II, in this country, scientists maintained a kind of noble impecunity (with some exceptions), largely independent of government money and control. As late as 1939, the federal government allocated only about $50 million for research, a small fraction of what would be spent after the war. Receiving little from the public sector, science also gave little: it did the research, and left the consequences up to society. There were a few small organized political activities, including a group supporting Hoover for President. (Hoover, an engineer, favored increased spending for scientific research.) And for about ten years after World War I there was a wave of interest among scientists in nonpartisan popularization of science, resulting, for instance, in the establishment of Science Service in Washington, D.C., and in the visibility of a few notable scientists, such as Robert Millikan.

But, by and large, attempts to influence policy, rather than to popularize science itself, were informal, individual, and infrequent.

The advent of the atomic bomb forced science and public to find a new working relationship. At a practical level, atomic energy research had turned physics into a much more elaborate, and expensive, undertaking. The Manhattan Project had involved most of America's top physicists in a massive, government-funded research effort. To continue the same quality of research, physicists would continue to need money and equipment such as only the government could afford. They would be dependent for funding and for freedom of research on the goodwill of politicians and the public.

On a more idealistic level, the bomb left some atomic scientists with a new-found social awareness, a sense of responsibility for the consequences as well as the causes of technology. The new feeling seemed to infect particularly those most responsible for the bomb's creation, men of the caliber of J. Robert Oppenheimer and Edward Teller.

The new sense of responsibility also came to have other components, including a feeling of obligation to report to the public the results of scientific research, since the taxpayers were paying the bills. And at times it was encouraged by a sense of mission, a desire to communicate the excitement of research, particularly to the young, the potential future scientists. Some scientists also felt that their profession gave them special attributes, such as a moral innocence, respect for facts, and familiarity with technical subjects, attributes which they should be willing to lend to the public policy debate.

In some ways the public encouraged this sense of mission, because its image of the scientist was likewise romanticized. Studies show that the public viewed scientists as highly credible sources of information, objective, methodical, open-minded, dedicated. Even congressmen pictured scientists as superior beings with special endowments, working for the joy it elicited rather than for money. The late scientist and philosopher Jacob Bronowski reflected the public's enthusiasm in 1956 when he wrote, "By the worldly standards of public life, all scholars in their work are oddly virtuous. They do not make wild claims, they do not try to persuade at any cost." Such a receptive audience would be difficult for scientists to resist.

Both the practical need for money and the moral sense of responsibility spurred scientists to get out of the laboratory and interact with government and public. And their new prominence from the Manhattan Project provided the opportunity. In keeping with their own elitism, most chose to explore channels at high levels of government, providing advice for federal officials rather than a general public.

At first, many government-scientist contacts were informal, as they had been before the war, but the system became more and more structured. There was a gradual institutionalization of both the "government of science," to handle funding of research, and of "science in government," to incorporate scientific advice into the policy-making process. Success in World War II legitimized the system of contracts as a funding mechanism and encouraged government-of-science organization along the lines of the wartime Office of Scientific Research and Development. Funding remained largely up to the Defense Department, with scientists as advisors and consultants, passing judgment on research proposals and suggesting further directions for research. By 1959, according to one estimate, there were forty-seven separate federal agencies devoted to research and development, each with its adjunct of scientific advisors. The institutionalization of science in government, which channeled scientific advice on policy issues, developed somewhat later, particularly after Sputnik, when special offices for science policy advisors were created in the White House. The advisory system became a complex system of temporary and permanent committees, an "establishment" of scientists involved in governmental decision-making.

The system was comfortable for scientists: money was plentiful, rising to levels over $15 billion by the mid-1960s, and social responsibility could be discharged in an easy, elitist way. And society got what it paid for: new weapons, satellites, technology for a race with the Russians.

The new equilibrium was easy, but it turned out to be unstable, as more and more problems that had technological components began to plague society. In particular, two "crises of confidence" within science affected the balance: scientists' loss of confidence in the public's continuing goodwill, and their loss of confidence in their own representatives in the Washington advisory system.

The first crisis in confidence is elusive. It is usually described as a situation in which the public lost confidence in science, but it might equally be viewed as scientists losing confidence in the public. It is not clear that the general public did in fact become disenchanted with science, but scientists and science-watchers at least thought they did. William H. Pickering, head of Cal Tech's Jet Propulsion Lab, lamented in 1974, "Science and technology changed almost overnight from hero to antihero."

Scientists based their concern on several trends. First, disillusioned opinion leaders in the media seemed to be blaming technology for a number of global problems, from environmental deterioration to totalitarianism. In the 1940s and 1950s, science had seemed a panacea, a solution to problems from hangnails to cold war. "Science" got the credit for producing a bomb to stop Hitler, then better bombs to keep abreast in the cold war, then satellites to catch up with the Russians. But ultimately science could not meet the public's, or its own, rising expectations. By the 1960s, problems were evident for which science had no answers. Instead of curing problems, technology was blamed for causing them: more virulent and widespread pollution, dirtier and more dangerous wars, greater disparity between the have's and the have-not's. In the scramble for federal funds, scientists also began to look self-serving, concerned with their own well-being rather than that of the larger society. Columnist Charles McCabe, for example, remarked, "I've gotten to the point where, when I see the words 'scientist claims' in a news story, I assume that the statements to follow will be high buncombe, tied in some way to the preservation of the system and of the values and paychecks of scientists in general, and this baby in particular."

Scientists could not avoid being affected by the bitterness of some critics, who seemed to be overreacting to the earlier era of naive goodwill toward science. In place of the old myths, with science as panacea, were substituted new myths, with science as powerful juggernaut. Sometimes these new myths were even contradictory; science was alternately viewed as an ineffectual academic exercise of no relevance to modern problems, and as an evil menace at the root of current crises. Man's walk on the moon, while a technological triumph, was criticized as a case of grotesquely misplaced priorities, an "arrogant piece of conspicuous consumption." If sustained technological effort could put man on the moon, why not direct that effort toward feeding man on earth?

Scientists were especially affected by the influence that public disenchantment appeared to have on the political scene, and thus on their freedom and their budgets. Describing the changing public attitude, Wolfgang K. H. Panofsky, head of the Stanford Linear Accelerator Center, reflected in 1971 that science was caught in a nutcracker, under attack from the right and the left. Forces on the right wanted research to show a better economic return; they wanted more practical results for their "science buck," as President Johnson once phrased it. Forces on the left added that the results should have social relevance, in areas like urban renewal, population control, and race relations.

Congress was thought to be reflecting the new antiscience mood when it began tightening both funds and restrictions for scientists in the late 1960s. For the first time in more than two decades, the allocation in 1969 for research and development was less than the previous year, falling from about $17 billion in 1968 to about $16 billion the following year. In the meantime, science had grown increasingly dependent on government spending for research, until almost two-thirds of funds for basic research and technological development were coming from federal agencies. As the flow of money slowed, the job market for scientists tightened, research possibilities narrowed, and applications increasingly took precedence over "pure" science. "My theory is that governments have paid up their mortgages to science," British molecular biologist Sydney Brenner has remarked. "After the war they took out a 25-year mortgage to pay back for the atom bomb and for everything else science had done through the war." By the 1960s, the mortgage was paid up, Brenner explains, and people began questioning the role of pure science. Looking at it another way, other observers have described the situation as a normalization of relations between science and government: science, as part of the knowledge industry, is being asked, just like any industry, to deliver when it gets federal money.

In 1972, the executive branch also took a hard line on "science," when President Nixon dismantled most of the formal structure for scientific advice in the White House. The position of Science Advisor, organizationally (but never politically) at a level with top foreign advisors and domestic aides, was relegated to the National Science Foundation, as one of the director's many roles. The elite President's Science Advisory Committee (PSAC) was disbanded — "excused," as a White House bureaucrat put it. The Office of Science and Technology (OST) was also abolished and its functions transferred to the National Science Foundation. Epitomizing what seemed to be the White House mood, a high official told *Science* magazine, "Science isn't a superior thing of itself which we have to keep on a pedestal. . . . If science has been downgraded, it is because it has been downgraded by society — and by the people making the reorganization plan."

In fact, the trends may not be as serious for science as they first appear. Polls do not corroborate a drastic change in public attitude. Sociologists Amitai Etzioni and Clyde Nunn studied the existing data and concluded in 1974, "the major shift . . . was not from great enthusiasm to great hostility, but from 'great confidence' to 'only some confidence' — a middling shift by all accounts." In fact, they find, science as an institution has fared better than most in an era of general distrust toward social institutions. There is also general agreement among poll-takers that the proportion of Americans actually hostile toward science is small, probably less than ten percent; positive attitudes are expressed by the majority. Scientists can also take hope from a recent survey of Californians, suggesting that the public makes a distinction between science and technology. While the public is indeed distrustful of technological developments that result from applying science, it is sanguine about scientific work per se.

Observers believe Congress is cool toward science, but following a slump between 1969 and 1971, federal budgets for research and development have been rising, albeit not always enough to keep up with inflation. Also, while Nixon was frosty toward the scientific community, relations thawed as soon as President Ford took office. Word was soon out that Ford would be receptive to the idea of reinstating a science advising system in the White House, and Vice-President Rockefeller voluntarily appeared at congressional committee hearings on behalf of the bill to set up an Office of Science and Technology Policy. "It was the first time in the history of the United States," Rockefeller noted, "that a Vice President has testified before a Congressional committee — and it was in the interest of science." Scientists were promised an even warmer reception from the White House with the election in the fall of 1976 of Jimmy Carter, who had had training in nuclear engineering.

What scientists and observers have interpreted as widespread "antiscience" is perhaps a collection of related but less drastic trends. One of these trends is the inevitable change in priorities in national policy. In the early 1970s the public's attention was diverted from the arms race and space exploration to cancer cures, energy alternatives, environmental improvements. The public still looked to science to solve its problems, but the problems changed. In the process, interest shifted from physics, associated with war technology and abstract research, to biological and behavioral sciences, related to domestic concerns. Physics, formerly at the top of the pecking order, felt the chill as it was forsaken, but the situation was not a crisis for science as a whole.

One of the recent beneficiaries of shifting priorities has been energy research. Indicative of the continued faith in science, Nixon argued in a November 1973 television speech on the energy crisis that, just as American spirit and scientific prowess had come through when we needed an atom bomb, and when we wanted a man on the moon, it could correct American fuel shortages. Toward that end he proposed allocating more research and development funds than the Manhattan Project had used.

In this sense, the government has always demanded relevance from science. According to science historian A. Hunter Dupree, government and public have consistently looked for practical results from basic research: "Even when one argued for basic research as an activity removed from useful application, the unspoken addition to the argument was that basic research would pay off in the future. Hence the argument for basic research was not a recognition of the cultural value of knowledge or of the necessity for healthy universities, but an argument merely for deferred practicality."

Whether or not scientists were justified in their antiscience fears, the public change in emphasis, the demotion of science advisors, and especially the cuts by Congress, darkened the mood of the scientific community and convinced scientists that their public image needed a face lift.

While losing faith in the public, scientists were also losing faith in their own leaders, creating the second "crisis in confidence." The traditional advisory system had had a number of problems. Many of the proceedings and studies scientists had conducted for the executive branch had been kept secret, leaving the rest of the scientific community wondering where their representatives stood, and why. If reports from scientific panels were unfavorable to the administration's position, they could be classified and pigeonholed; this apparently happened during the supersonic transport (SST) battle. If reports were made public, they could also be presented so as to be misleading to press and public. Unwanted reports could be delayed; Frank von Hippel and Joel Primack described in *Science* this example of bureaucratic foot-dragging:

In 1966 a report by an independent laboratory under contract to the Department of Health, Education, and Welfare indicated that 2,4,5-T (2,4,5-trichlorophenoxyacetic acid), a popular weed and brush killer, causes birth defects. This report was repeatedly sent back for "further study" for 3½ years until it finally became public as an indirect result of a Nader investigation. In the meantime, enormous quantities of this chemical were used in the defoliation of about one-eighth of the area of South Vietnam.

Outside scientists, who were not privy to the classified information on which reports were based, were helpless; if they objected, they could be dismissed as know-nothings.

It was also obvious that some scientists in Washington had a conflict of interest. Scientists often advised the very agencies that funded them; one study showed that 86 percent of the members of ABM advisory panels in the Defense Department got their research support from that department. The trouble with having scientists advise funding agencies was that, as Alvin Weinberg put it, "judge, jury, plaintiff, and defendant are usually one and the same." Also, too much power in Washington seemed to be concentrated in a few influential scientific advisors, a self-perpetuating, self-selecting elite, in which new members were carefully screened for political pliability. These "consistently influential" advisors numbered only two hundred to four hundred. Only the "tame cat scientists" were heard, says Philip Abelson, editor of *Science* magazine. After less than twenty years, the advisory system was showing signs of old age.

The effectiveness of the scientific advisory system diminished especially in the last half of the 1960s. James D. Watson, biologist and author of *The Double Helix*, attributes the change to the Vietnam War. When some university scientists became outspoken critics of the war, the whole community lost credibility with Johnson, Watson feels. A vacuum in scientific leadership developed. And it still exists, according to Watson, because potential leaders lost confidence in their own ability during this period, and dropped from the Washington scene. In essence, leaders formed by World War II were "de-formed," in Watson's words, by the Vietnam War. According to John Gardner, chairman of Common Cause, there were also other deterrents to leadership roles; there exists a kind of "antileadership vaccine," he says, made up of discouragements created by our mass society, by the bad image of politicians and administrators, and by the narrow training received by today's scholar, scientist, or professional.

Edward E. David, Jr., the last science advisor before the 1973 change, feels scientists lost influence also as a result of the shift in emphasis from military and space to health and social problems. The new decisions, he says, had a larger "human" and moral component, in which government was less receptive to a scientist's advice. That scientists were less effective in social areas may have been a factor in the decline of the President's Science Advisory Committee, for example. The achievements of PSAC, says former presidential science advisor George Kistiakowsky, were less impressive in civilian areas than in "national security" matters. In Kistiakowsky's analysis, as PSAC diversified into domestic areas, its credibility and cohesiveness were weakened. PSAC had also been instrumental in setting up offices for technical input at all levels of the executive branch, and the infusion of scientists and engineers detracted from the influence of PSAC. In essence, a former PSAC member says, PSAC "sawed off the limb on which it sat." PSAC was also a victim of powerful internal politics, former members point out. While Eisenhower took a personal interest in his science advisors, later science leaders ran afoul of Robert McNamara, Henry Kissinger, and other advisors jockeying for influence with the President. Gradually PSAC and OST tumbled further down the White House hierarchy, until they were dispensed with altogether.

The Outsiders

Whatever its cause, the vacuum in scientific leadership in Washington occurred at a time when scientists felt they needed a strong voice in policy, both to counter "anti-science" moves on their own behalf, and to help with increasingly serious technological problems. The situation was grim enough to loosen an unusual number of scientists from their laboratories, in spite of strong pressures from the scientific community to concentrate on basic research. Motivated by practical need to foster funding, and an ideological commitment to social issues, these scientists increasingly engaged in public activities — writing articles and letters-to-the-editor, giving speeches, organizing movements in professional societies, working in political campaigns, taking litigation to the courts, lobbying in Congress. In particular, they sought direct contact with the public, holding an implicit belief in the workings of the democratic process and the ability of the individual citizen to exert influence on policy.

Consciously or unconsciously, these scientists functioned as an alternative to the advisory system, working to counteract its failings and inadequacies. Where advisors were "insiders," they were "outsiders." Where the advisory system was elitist, outsiders strove to keep organization unstructured and democratic. Where advisors often took nonpartisan postures to avoid jeopardizing their positions, outsiders were usually frank proponents of one side of policy issues. Where advisory panels tended to be secret, outsiders worked to bring issues into the open. Where advisors tended to be the older, established scientists, outsiders were more likely to be young and untried. The advisory system was "establishment," the outsiders antiestablishment.

As a counterforce to the advisory system, outsider activity tended to thrive at times when feeling was strong in the scientific community that the inside advisory system had failed them on a key issue. A precursory case was the atomic scientists' movement following World War II, which evolved around the battle over whether civilians or the military would control future atomic energy policy. It was a battle from which outsiders felt their inside leaders had retreated. In 1945, the wartime administrators of the government research labs, who gradually became the inside advisors, supported the Truman administration's May-Johnson bill, which would have given the military control over aspects of atomic energy. Atomic scientists were disillusioned with their leaders and the whole coterie approach to influencing policy. A Chicago physicist wrote to a colleague, "I must confess my confidence in our leaders Oppenheimer, Lawrence, Compton, and Fermi, all members of the Scientific Panel advising the Interim Committee . . . is shaken." The disillusioned outsiders initiated a massive lobbying campaign aimed at official Washington — letters, visits with local officials, personal missions — and a publicity campaign aimed at the general public —

speeches, pamphlets, radio, magazine articles, and finally a month of barnstorming in April 1946. Their reaction induced the substitution of the McMahon bill, which, with the scientists' support, resulted in the establishment of the civilian-controlled Atomic Energy Commission.

The pattern has remained the same, with surges of outside activity over arms control, Vietnam, environment. The ABM controversy was generated partly by discontent with the secrecy of the advising process, a feeling that no one knew where the insiders stood. And as the Vietnam War worsened in the Johnson administration, even insiders began working on the outside to try to change the course of the war. In 1970 Kistiakowsky and several other key insiders opted to give public support to the congressional move to limit the President's powers to make war in Southeast Asia. Said Kistiakowsky, who had had the top scientific insider post under Eisenhower (1959–1961), "Most of us had been members of the establishment and believed that we could accomplish more by working through channels. . . . But now we have reached the conclusion of desperation that this is impossible. . . . Our prime and most intense concern is to assist in building up a mass movement all over the country to put pressure on Congress and the administration to end the war."

The problems have persisted and proliferated, and with them the outsider scientists. Most outsiders have a university position for a platform, but a few have quit research altogether to become full-time writers or organizers. Peter Metzger, a former Columbia University biochemist, is now devoting full time to muckraking and newspaper writing in Colorado, where his efforts as a leader of the Colorado Committee for Environmental Information have made news, uncovering such local environmental hazards as inadequate nerve gas storage at the army's Rocky Mountain Arsenal and plutonium dangers at the Rocky Flats AEC plant. Some outsiders are rugged individualists; some work through organizations. And organized groups are as diverse as their members, ranging from the Scientists' Institute for Public Information (SIPI), headed by Barry Commoner, to the Federation of American Scientists (FAS), a lobbying group now revitalized and directed by Jeremy J. Stone, or to Science for the People, a radical group formed in 1969 and dedicated to politicizing the professional societies and universities. Many outsiders are Nader-like "public interest scientists," described, for example, in Joel Primack and Frank von Hippel's book, *Advice and Dissent: Scientists in the Political Arena.* The new congressional fellows, scientists funded by the American Association for the Advancement of Science, the American Physical Society, and other professional societies, who work as technical experts on congressional staffs, could also be considered outsiders. A few scientists are both insiders and outsiders. Drell viewed his activities in PSAC as "inside," his membership in FAS as "outside," the one organization operating as a kind of balance for the other in Washington politics. Geneticist Joshua Lederberg has both served as a top-level advisor and written a column in the *Washington Post.*

Politically, outsiders can be divided, as U. C. Berkeley physics professor Charles Schwartz has suggested, into "political radicals, or philosophical idealists" and "political liberals, the pragmatists." As a radical, Schwartz sees his role as opposing the centralization of military, economic, and political power, which science and technology have helped to maintain and extend; he would "put the resources of scientists at the service of those large numbers of people farthest away from the centers of control, rather than allowing those already in the seats of power to increase their power through the use of science." Schwartz examined, for example, the names of the thirty-one scientists who had been members of PSAC from 1967 to 1970 and found that, while they were predominantly from universities, "one-third of these academics have sat on the boards of directors of corporations doing over $100 million annual business in technological fields."

More conservative outsiders are willing to compromise with the establishment and the press in order to further specific objectives, such as elimination of military use of defoliants in Vietnam, or better safeguards in nuclear power plants. Bitterness between radical and pragmatic scientists is often apparent, as when squabbles arise over policy for conducting demonstrations at professional meetings.

Despite their diversity, the outsiders share a political approach based on faith in the people, the final decision-makers, and the ultimate beneficiaries of their efforts. Epitomizing their approach, Paul Ehrlich explained in *The Population Bomb*, "I am becoming more and more convinced that the only real hope in this crisis [population] lies in the grass-roots activities of individuals. We must change public opinion in this country, and through public opinion change the direction of our government."

Their dedication to reaching the general public makes the outsiders by definition somewhat visible, and a number have become known, or notorious, in local areas and for short crises. At the tip of the iceberg, a few outsider scientists have become relatively widely known nationally — today's visible scientists.

Glenn Seaborg: An Insider

One way to see the difference between today's visible scientists and the traditions from which they come is to contrast an example of the new breed with one of the old. One of the best examples of the postwar "establishment" scientist is Glenn T. Seaborg. A Nobel Prize-winning chemist and for ten years the chairman of the Atomic Energy Commission (AEC), Seaborg is a visible scientist of the "old" type. He is as outstanding and unusual as a Paul Ehrlich or a Margaret Mead, but in a totally different way — the established, accepted way for scientists of his time.

To the mind of the traditional scientist, Seaborg does everything right. His storybook career reads like a dream come true for scientists of his era. Although he had to work his way through school, he was the valedictorian in his class at a high school in the Watts district and was elected to Phi Beta Kappa in his junior year at the University of California, Los Angeles.

From a family of Swedish machinists, he fashioned his own career, and was encouraged by high school and college teachers to go into science. He chose the graduate school with the best reputation and staff in chemistry, the University of California at Berkeley, completed a Ph.D. with dispatch (1934–1937), and, as every young chemist must have fantasized, was invited by "the great G. N. Lewis," physical chemist at Berkeley, to be his personal research assistant. As a graduate student, Seaborg also happened "almost by accident" into an important new field in chemistry: Berkeley physicist Jack Livingood invited Seaborg to join him in the discovery of radioisotopes. From 1937 to 1940 Seaborg and colleagues published over forty papers on new isotopes, including iodine-131, used as a radioactive tracer in the treatment of thyroid disease. Seaborg's mother was one of the many people whose life was later saved by the iodine-131 technique.

The isotope work led Seaborg to the study of transuranium elements. In 1940 E. M. McMillan and Philip H. Abelson announced the discovery of an element heavier than uranium (atomic number 92), then believed to be the heaviest element in existence. The new element, named neptunium (atomic number 93), was made by bombarding uranium with neutrons in the giant cyclotron at Berkeley. When McMillan was called away for war-related activities, Seaborg inherited his work, and in 1940 took part in the discovery of a new transuranium element, plutonium (atomic number 94). Almost immediately, in 1941, a second isotope of plutonium, plutonium-239, was produced, then in 1942 an isotope of uranium, U-233, both fissionable and thus important potential fuels for atomic fission reactions. Seaborg was, not surprisingly, invited to work on the Manhattan Project, and in 1942 went to the University of Chicago Metallurgical Laboratory to head the section in charge of developing a chemical method for separating plutonium from uranium in order to provide fuel for the hoped-for atomic bomb. From 1944 to 1958, he also participated in the discovery of eight additional transuranium elements, and formulated the "actinide" concept to predict and explain the role of the transuranium elements in the periodic table.

His career "made" by the discovery of plutonium in 1940 (he was twenty-eight), and capped by the Nobel Prize he shared with McMillan in 1952 (he was thirty-nine), Seaborg channeled his energies more and more into administrative and advisory duties. For a scientist of stature with major research contributions behind him, prestigious advisory and administrative positions were the accepted and approved next steps. And Seaborg's positions from the beginning were prestigious. After the war, in 1946, he was appointed by President Truman to the first General Advisory Committee of the Atomic Energy Commission, whose members had been the leaders of the bomb project: Enrico Fermi, J. Robert Oppenheimer, I. I. Rabi, James Conant, Lee DuBridge, and others. Seaborg, at thirty-four, was the youngest member, and impressed that he was included.

Although Seaborg says that he found the bimonthly trips to Washington distracting from research and that he was "extremely relieved to get off the GAC" in 1950, in 1952 he agreed to be University of California's Faculty Athletic Representative to the then Pacific Coast Intercollegiate Athletic Conference. The position seemed an unusual one for the chemist, but he points out that it made sense, given his "lifelong interest in athletics." The job caught him up in a period of turmoil over widespread rule violations within the conference, and the conference ultimately broke up. Seaborg's adeptness at handling the controversy and the setting up of a new conference, the Athletic Association of Western Universities, was not overlooked. The job turned out to be a steppingstone to the chancellorship of the University of California at Berkeley, and from there to the chairmanship of the Atomic Energy Commission. (His administrative abilities had attracted attention as early as the Manhattan Project.) He survived two changes in administration and resigned from the AEC in 1971, when he returned to Berkeley to be associate director of the Lawrence Berkeley Laboratory, head of the Nuclear Chemistry Division, and University Professor of Chemistry for the University of California campuses. Seaborg is careful to point out that, even while serving as chairman of the AEC, he kept up with developments in his field, and that he now once again combines administrative duties with teaching and research. He teaches a full schedule (freshman chemistry laboratory, seminars, graduate student research), and lectures on all nine University of California campuses; his research has involved him in discovery of a new transuranium element, his tenth, element 106. He has also continued to be active in the American Chemical Society, and was elected its president for 1976, an important year for ACS, Seaborg feels, because its centennial coincides with the United States' bicentennial.

Although Seaborg expressed surprise at each new development and reluctance to take on each new administrative task in his career, actually it worked out to be almost a model for the proper, ordered scientific career of the day: early research breakthroughs, contribution to the bomb project, more research coupled with advising in Washington, and full- and part-time administration. Noting the orderly pattern, Seaborg wrote in an autobiographical sketch:

In retrospect, I can see my stepwise progress, how each phase of my career prepared me for the next. The extraordinary, I called it magic, Berkeley atmosphere of the 1930s offered unparalleled opportunity. My close association with G. N. Lewis showed me firsthand how the world's best chemist worked and thought, and gave me, a young chemist rather lacking in self-confidence, the confidence that was essential to future progress. The association with the Radiation Laboratory, and the amazing Ernest Lawrence, made possible the discoveries with which I was involved before the war. This experience made it possible to carry out my subsequent responsibilities at the wartime Metallurgical Laboratory of the University of Chicago. These experiences, in turn, made it possible for me to carry out my research role in the Radiation Laboratory during the productive period 1946 to 1958. All of this experience, and especially my experience as the Faculty Athletic Representative to the PCC.[Pacific Coast Conference], prepared me for the Berkeley chancellorship. And the administrative experience gained in the latter position, together with all of my research experience, prepared me for the chairmanship of the AEC.

Even Seaborg's personal life seemed orderly: he married Helen L. Griggs, then secretary to the Berkeley Laboratory's eminent director, E. O. Lawrence, just as his war work was beginning;

they decided to have, and have had, six children, whom they raised in the prosperous San Francisco suburb of Lafayette, except for the ten years in Washington. Seaborg works hard but takes time out for exercise and recreation and a good eight-hour sleep each night.

In contrast to most of today's visible scientists, Seaborg has been very much a member of the scientific establishment, very much an insider. By the time he was forty-eight and new chairman of the AEC, he was considered by science observers to be one of the "elder statesmen of science." While he has played an award-winning role in the popularization of science, he has confined himself to noncontroversial, apolitical topics and traditional channels in his contacts with the public.

Seaborg's decision to make public, nontechnical speeches was a hard one, although in retrospect, he feels, the right one. "By the time I'd gotten through graduate school, I could speak on a scientific subject without any nervousness," he says. "But at the beginning, twenty years ago, I found speaking on nonscientific, sociological topics rather awesome and difficult, and I was nervous." All it took was a little practice, and by the time he was giving press conferences as athletic representative, he positively enjoyed meeting with reporters, parrying their questions, maintaining everyone's good humor.

Demands for public speeches increased as Seaborg became chancellor and AEC chairman; at times he averaged an invitation a day. Since his retirement from the AEC he has given fifty to a hundred speeches a year to national and international audiences of young people and adults, on topics such as "science, technology and the quality of life," and "energy and environment." He is also president of the Board of Trustees of Science Service, which runs science fairs, Science Talent Search, and other science popularizing activities, and he is a member of the Board of Directors of National Educational Television and Radio Center (NET).

Seaborg's approach in speech making and popularization of science is the same as in administration — cautious when it comes to controversy — and this perhaps more than anything else sets him apart from the new, outspoken, outsider scientists.

He makes a policy of trying never to anger anyone, writer George Boehm has observed; Boehm relates this unusual story:

One day . . . while he was on a prolonged speaking tour, Seaborg was suddenly overwhelmed by revulsion at the prospect of more rubbery chicken and cold peas. He had always been a steak-and-potatoes eater, and he decided that he had to have a steak. Telephoning ahead to the chairman of the next night's meeting, he announced that he was arriving on the seven-o'clock train. Then, guilefully, he boarded an earlier train. He arrived in plenty of time to enjoy a steak dinner, after which he hurried back to the station to be met by the welcoming committee. At the banquet that followed, he complained of dyspepsia and cheerfully passed up the chicken.

Nowadays, he turns down most speaking invitations; he accepts only when he feels that by refusing he'll "make more enemies than is worthwhile." Using this "process of elimination," he still ends up with too busy a speaking schedule.

By the end of his AEC years, Seaborg had become, according to *Time* magazine, "a legend in Washington for his ability to duck controversy." As an example, *Time* continues, "During the intense debate over whether the United States should build an H-bomb, he managed to retain the friendship of both Robert Oppenheimer and Edward Teller. On the one hand, he agreed with Teller that the bomb should be built; on the other, he so qualified his support that even Oppenheimer, the project's chief opponent, could hardly take offense."

Seaborg also weathered the infighting during the breakup of the Pacific Coast Conference without losing the friendship of any of the other representatives. Later, his term as chairman of the AEC saw bitter fighting over the safety of the AEC's radiation standards for nuclear reactors, but Seaborg maintained his equanimity throughout. Still later Seaborg emerged unscathed from an unexpected battle over his candidacy for presidency of the American Association for the Advancement of Science in 1970. He was still chairman of the AEC, and although he and the White House knew he was planning to resign before he would be taking office as AAAS president, it appeared to critics as if the two posts would be simultaneous and present a conflict of interest. Members who felt the AAAS should be seeking a socially responsible stand on uses of nuclear power argued that Seaborg, as chairman of the AEC, would not and could not take an objective part. A casualty of the fight was Daniel Greenberg, who resigned from AAAS's *Science* magazine staff when he was restrained from reporting the dispute before the election. Seaborg was taken aback by the controversy but kept his composure, explaining to *Science* that it would have been "embarrassing or awkward" to turn down the post for the third time, and "I really want to do what's right for AAAS." Why hadn't he let AAAS leaders know of his plans to resign from AEC? In a 1973 interview, he said that if they had only asked, "I could have given them a hint." Instead, after much deliberating and consulting with representatives on all sides, Seaborg simply left his name on the ballot and was elected, serving as AAAS president in 1972.

The only people he antagonizes, then, are those who do not like his neutrality. He is either "calm" or "cold," "diplomatic" or "indecisive," depending on the point of view. Not surprisingly, the new politically oriented, strong-minded visible scientists are not likely to be his admirers. Some of his critics over nuclear safety standards feel he served as a "marvelous front" while at the AEC, a high-prestige symbol, providing no leadership. Peter Metzger, in his book *The Atomic Establishment*, says that the ten-year period in which Seaborg was chairman of the AEC is regarded by insiders as a period of "no chairman at all." "He would like to be right, no matter what happens," explains John Gofman, a nuclear power critic, who used to be a student of Seaborg's.

Seaborg antagonizes the outsiders because insider fence-straddling is just what they are against. Expressing outsider frustration with the politics of establishment scientists, Metzger wrote:

No greater symbol of this abstract, docile, apolitical and yet at the same time arrogant stance exists than . . . Dr. Glenn T. Seaborg. His voice is gentle, his actions noncontroversial, and he plodded quietly on, from speech to speech, promoting atomic energy, and

AEC, and his own pet interests. . . . He had been chairman of the AEC through Presidents as diverse as Kennedy, Johnson and Nixon. "President Nixon's not much different," he said in a 1970 interview. "The overall thrust continues."

Perhaps it is partly a reflection of a changing mood in science news that, while Seaborg was generally praised by journalists ten years ago ("enormous political skill," "an adroit reconciler of divergent personalities"), the media are now cooler ("his ability to duck controversy," and "celebrated sang-froid").

The media seem to be setting new standards for visibility, gravitating toward the active, politically involved scientists. Seaborg and other insiders were the dominant visible scientists of the 1940s and 1950s. Today, with a dearth of leaders on the inside, and a wealth of science-related issues on the outside, the outsiders have come to predominate.

Barry Commoner: A Guerrilla

The visible scientist most conscious that his new role makes him an outsider is Barry Commoner, a biologist at Washington University in St. Louis. Dubbed the "Paul Revere of ecology" in a *Time* magazine cover story, Commoner gets his visibility from environmentalist books (*Science and Survival, The Closing Circle, The Poverty of Power*) and from frequent lectures and television appearances as a leading environment expert and spokesman. Acknowledging his "outsider" status, he said in an interview, "You understand that there is an Establishment . . . and you either have to decide that you are going to bend yourself to do the right thing and conform, not upset people — or if you don't, you just turn your back on it."

If Seaborg does everything right, Commoner does everything wrong. More than "turning his back" on the Establishment, Commoner defies it. He has made being "outside" his life style. He tilts with his university administration, his own department, establishment biologists, the political status quo, and the Washington scientific advisory structure.

Underlying Commoner's actions is a belief that the public needs to be informed directly and fully by scientists on the technical aspects of social issues. He chose fallout and disarmament in the 1950s, environment in the 1960s and 1970s. What has come to be known as the information movement, spearheaded by Commoner and Margaret Mead, is based on a philosophy that the scientist should not divorce himself from issues raised by his work; nor should he confine his efforts to advising officials in Washington; nor, at the opposite extreme, should he involve himself as a prophet, commenting both on the technical and value aspects of the issues. Instead, the scientist should use his special background to provide citizens with facts, which they then use to make informed decisions. The scientist fulfills his obligation by alerting the public, supplying the facts, "delivering to the public the grist for the mill of their conscience"; the public is willing, able, and obligated to do the rest. Thus Commoner and the information movement stand at odds with the traditional view that scientists belong in the laboratory, and they also differ from many other scientists who have ventured outside the laboratory. Many, of course, question whether Commoner can, or does, divorce "fact" from "opinion," as he claims in popular lectures and writing. According to Harvard University's Roger Revelle, "he has tended to push a kind of lawyer's brief rather than an objective examination of all the evidence."

Commoner's arm of the information movement had its beginnings in a very "established" organization, the American Association for the Advancement of Science (AAAS). Warren Weaver, mathematician and director of the division of natural sciences at the Rockefeller Foundation, was convinced that the AAAS needed activists and urged Commoner to become involved in the organization. Commoner started at the bottom rung, as secretary of the botanical section (1954–1960). Later, as chairman of the section (1960–1961), he began introducing into the agenda discussion of problems relating to social responsibility. He met Margaret Mead, who was also working within the

AAAS, and together they organized ad hoc meetings, and finally in 1958 persuaded the AAAS to institute a Committee on Science in the Promotion of Human Welfare, Commoner chairman (until 1965), Mead board representative. In 1958 the committee held a symposium on fallout, where the philosophy of providing facts directly to citizens was put forward. In an article based on his speech at that symposium, the first written statement on the information approach, Commoner said:

What we need now is to marshall [sic] the full assemblage of facts about fallout, their meaning and uncertainties, and report them to the widest possible audience. This is not an easy task. It is much simpler to publicize conclusions alone, and have them accepted not because their factual origin is fully understood but because they carry the authority associated with science.

It seems to me that we dare not take this easy way out. Unless the public has sufficient information to provide a reasonable basis for independent judgment, the moral burden for the future effects of nuclear testing will rest on some smaller group. And no such group alone has the wisdom to make the correct choice or the strength to sustain it.

. . . a public informed on this issue is the only true source of the moral wisdom that must determine our nation's policy on the testing — and the belligerent use — of nuclear weapons.

Commoner proceeded to demonstrate the philosophy in action by organizing groups and giving public lectures. The St. Louis Committee for Nuclear Information (CNI) [later, as times changed, known as Committee for Environmental Information (CEI)] was formed that year. A general Scientists' Institute for Public Information (SIPI), founded in 1963 as an outgrowth of CEI, now incorporates CEI and about sixteen local groups like it around the country, and publishes *Environment* magazine, offspring of a publication started with CNI. Commoner, as chairman of the board of directors, has been considered the leader of SIPI. (Margaret Mead was president until 1973.)

In keeping with his philosophy, Commoner has also refused invitations to take positions in the insider advisory system in Washington. "I made the mistake of accepting one government position," he says, "and I regretted it." The position, in 1969, was on a Department of the Interior committee to study the SST's effects on the environment. Commoner found the experience "totally ridiculous" because he could not share what he learned with the public. After two meetings in 1969, he quit. He used to be invited to serve on other inside committees, but no longer. Insiders, in turn, have been scornful of the information movement, finding that it is not "strong enough meat" for anyone who wants to have real political impact. In fact, at the beginning, Commoner recalls, "the establishment figures looked on this as an insignificant kind of thing; of course the obvious way for the scientist to exercise his social responsibility was to get an appointment in the White House."

Along with SIPI, Commoner has been active politically — "involved," he says in his hyperbolic style, "in every left-wing political activity in the academic world." He was one of the organizers of the national teach-in on Vietnam, for example, during the teach-in era. While such activities were hardly unconventional, they were outside the usual scientist's realm, and they did cause trouble with the university administration. "If I hadn't had tenure," Commoner says, "I would have been out on my ear." The university board asked the administration, Commoner has been told, to "shut me up" in a number of instances, and a former chancellor once ordered the university news bureau not to give Commoner any publicity. He believes some of the trouble arose from the fact that the head of Monsanto Company, which had been accused of manufacturing environmentally detrimental chemicals, was chairman of Washington University's board.

A penchant for taking "the road less traveled" has been with Commoner for some time. His first job, after a stint with the navy during the war, was as associate editor for *Science Illustrated*, a New York magazine that popularized science by featuring large pictures, sensationalized stories, and a pretty girl on the cover. "The magazine started out to do a serious job along the lines later made successful by *Scientific American*," Commoner explains. "I joined the staff on that basis and was one of a group on it that struggled, unsuccessfully, with the McGraw-

Hill management to keep on that path. [I] spent many anguished months in this struggle. . . ." He remembers, for example, a "huge battle, which I lost," over his article on venereal disease. The article ultimately appeared under the title "The Menace of Venereal Disease" and was accompanied by a picture of comely German girls lined up for VD tests. The text described the fate of American servicemen afflicted with venereal disease in Germany, offered information on tests and treatment, and provided a questionnaire to test the reader's knowledge. The pay was good, Commoner reflects, and, more important, the job gave him an opportunity, after four years in the military, to catch up on science, to review the literature, and to decide on a new path of research.

A year later, in 1947, the magazine was on the verge of folding and Commoner was ready for a university position, but again he took the independent approach. He turned down the idea of working at Yale University with Joshua Lederberg and Edward Tatum — two geneticists who later (1958) won the Nobel Prize for their discoveries about bacterial genetics. Instead, Commoner opted to strike out on his own, with a post as associate professor at Washington University in St. Louis. "And it turned out to be a very wise move, too," he says, "because I got away from the whole eastern establishment, where everybody was talking to everybody else; I thought my own thoughts out here."

At the same time he decided to do research on tobacco mosaic virus (TMV), although he had never worked with a virus before. The Rockefeller Foundation turned down his request for funds, concerned about his lack of experience with TMV. Instead, the foundation offered Commoner a fellowship to study under W. M. Stanley, a Nobel Prize-winner at the University of California, Berkeley, and the top man in the TMV field. Commoner turned it down: "I was pretty well set in the idea that it was better to strike out for yourself than to get forced into the establishment's pattern. . . . I'm sure that Stanley was — he's dead now — offended by my independence of him." Commoner studied TMV for a year on his own, and the following year got his Rockefeller grant. In 1953 his report of his findings at the AAAS meeting won him the Association's New-comb Cleveland prize, given to the young scientist who delivers the most noteworthy report at the association's annual session.

A second stream of research involved him in an investigation of the role of free radicals (special molecules with unpaired electrons) in the cell's metabolism. He organized a research group to detect these radicals by means of an electron spin resonance (ESR) spectrometer. Commoner, who is sensitive about the fact that public and colleagues alike are less aware of his laboratory work than his public prominence, calls it "pretty damn esoteric biophysics," and notes that he has coauthored about twenty papers on the subject. At first his ESR publications were not readily accepted. Commoner gets a kick out of repeating the story George Wald told him. When the first ESR paper came out, Wald, a Nobel Prize-winning biologist at Harvard University, had not heard of the phenomenon of electron spin resonance and thought Commoner had fabricated a far-out physical principle. "He went trotting, of course, to his physicist friends, and found that some of the basic work had been done at Harvard." The ESR work is now professionally recognized, according to Washington University physical chemist Samuel Weissman, a member of the National Academy of Sciences, and Commoner himself believes it is "cited as the pioneer research." But he points out that "as the ESR field has grown into its own little establishment, I have deliberately avoided becoming a part of it, preferring to move along my own path." All told, Commoner says, "I'm sure that if I hadn't sort of 'dis-established' myself in general, people would be paying more attention to the free radical work."

About 1966 Commoner dropped the TMV work, on which he had published about twenty-eight papers, and made environment his research as well as political interest, concentrating on the study of the nitrogen cycle. Commoner finds this research interest also unusual: "I tend very strongly to do idiosyncratic work. . . . I have a strong feeling that there is not much point in doing something that someone else can do."

One of Commoner's most startling departures from established doctrine has been a holistic conviction that DNA is not a "master molecule," as proclaimed since the 1950s by a majority of biologists (and lay public and press). The central dogma since about 1958 has been that DNA (deoxyribonucleic acid) controls the cell. Other factors, such as radiation and drugs, may influence the genetic mechanism in a random way, but they are not basic carriers of genetic information. Commoner maintains that, while information is indeed passed from DNA to the cell, protein also transfers information back to DNA in a cyclical pattern. For example, the enzyme DNA polymerase, whose structure is dictated by DNA, in turn influences the structure of new DNA in the cell. And some inheritance, Commoner adds, is not under the control of Mendelian genes, and cannot be explained by the standard one-gene-one-enzyme hypothesis. The whole cell, and not just DNA, then, is responsible for inheritance, and a holistic approach is needed to study the phenomenon. Reminiscent of the vitalists (although Commoner says his position is not vitalistic), Commoner asserts: "Biologists have confronted successively — like a nest of Chinese boxes — levels of complexity ranging from the ecosystem to the internal chemistry of the cell. The last box has been opened. According to the Watson-Crick theory, it should have contained the single source of all the inherited specificity of living organisms — DNA. It is my view that we now know that the last box is empty, and that the inherited specificity of life is derived from nothing less than life itself."

Such a deviant position has been ridiculed and ignored by most of the genetics establishment — although recently Francis Crick, recalling early reactions to the announcement that he and James Watson had discovered the structure of DNA, said, "Barry Commoner insisted, with some force, that physicists oversimplified biology, in which he was not completely wrong." Commoner says he began looking into the issue back in 1946, but worked on it for fifteen years before publishing anything, until he felt certain of his ground. He realized, "I was sticking my neck way, way out." He figured that his unorthodoxy would prevent him from being elected to the prestigious National Academy of Sciences. "And I knew people would be terribly upset by it, because it raised very complicated questions. I spoke on this theory in Russia, just at the time I published the papers here, and my Russian friends pleaded with me not to do it, because they said, 'We're just getting over Lysenko, and we've got a nice, simple theory that people will believe; don't make it more complicated.' But two or three months ago I got a letter from one of my Russian friends who said that he is beginning to think I was right." Describing his dilemma, at a 1968 conference, he said: "What I'm doing here is saying that the king is wearing no clothes. The question of differences in approach such as atomism and holism is ludicrously important, and yet there is a total lack of concern about this kind of question among practicing biochemists and molecular biologists — it's really a peculiar sort of cultural situation: the rigidity of the establishment extends to censorship, condemnation, inhibiting grants, and so on. There's a vast cultural superstructure here and I think it's important to keep in mind." Reflected Commoner in 1973, "If I hadn't offended the Establishment, my work would have justified my being in the National Academy a long time ago."

Whether Commoner deserves better recognition for his scientific contributions is difficult to assess. His strong personality and maverick stands win him enemies, emotional reaction, and partisan controversy, making it difficult to get objective assessments of the quality of his professional work. At Washington University, animosity is especially widespread. Graduate students used to refer to his presence in the department, back when he was associate professor, as "creeping Commoner-ism." Commoner was chairman of the botany department from 1965 to 1969, and fought bitterly a proposed merger between the biology and botany departments — a merger some biologists say was prompted by a movement in the Botany Department to get rid of Commoner as chairman. Commoner lost that round and was absorbed into the biology department, but the next round was more nearly a tie. When Commoner found out that his greenhouse was to be torn down to make room for a new biology building, he "threw a fit"; they built him a new greenhouse. Looking out his window as construction proceeded in 1973, he was still not happy, however, because the new biology building was going to block off two of the windows in his large airy office. "But what can you do about it? I have worked against opposition always, right?"

Over time, Commoner has nearly divorced himself from the biology department — "I'm exiled." Although he still holds a post as professor of biology, he teaches courses only occasionally and does not participate in academic committees. In 1966 the Public Health Service granted Commoner $4.5 million for a Center for the Biology of Natural Systems, funding which has freed him from university resources. He devotes much of his time to directing the center and to public lectures (his lecture fees go to SIPI). "I am increasingly separated from actual laboratory work," he says, having done little physical work at the laboratory bench since 1960. But he points out that he retains an active part in the research process, working with research associates and assistants in the design and interpretation of experiments and writing papers.

In spite of academic troubles, Commoner has been a clear favorite with the press. *Time*'s cover story on Commoner was one of its first forays into environmental coverage. Acknowledging that he is "sometimes aggressive and even abrasive," *Time* observed that "he is endowed with a rare combination of political savvy, scientific soundness and the ability to excite people with his ideas," and he is "revered as the voice of reason in a lunatic world." Reporters have appreciated particularly Commoner's articulateness. Says a St. Louis writer, "He makes Billy Sunday look like a speech problem." Commoner laments that when he gives even a scientific paper, people compliment him on how well he presented the material; "I wish they would say, 'What remarkable experiments!'"

While Commoner has a reputation for egotism among scientists and reporters, he wins points for not talking down to the layman. "Unlike some colleagues," Daniel Lang observed in reviewing Commoner's first book for the *New York Times*, "Mr. Commoner shows no condescension to laymen." On the contrary, Lang found the book, *Science and Survival*, "modest and temperate throughout," and praised Commoner's "instinct for fairness."

Commoner used to antagonize reporters, when he aggressively cultivated press contacts at AAAS meetings and CNI functions. "He would come up and buttonhole you," one science reporter complained, "and tell you he had a great story." ("There obviously was a time," Commoner responds, "when I thought that environmental stories were more important than most science writers did.") As his popularity and the popularity of the environmental issues have increased, his overtures have decreased; the press comes to him. He was nearly indispensable in the early environment movement. During the first heat of press environment coverage, environment reporter David Hendin observed, "If you get a shocking environment story that you want to write thoroughly, you have to call up Barry Commoner." On Earth Day, April 22, 1970, Commoner gave speeches at four universities in one day: Wellesley, Brown, Rhode Island University, and M.I.T.

The press also recognizes that it is dealing with an unusual kind of scientist. Summarizing Commoner's role in its 1970 cover story, *Time* wrote: "Barry Commoner is a professor with a class of millions — most of them real students, all of them deeply concerned about man's war against nature. At 52, the impatient microbiologist from Washington University in St. Louis has become the uncommon spokesman for the common man. He personifies the New Scientist — concerned, authoritative and worldly, an iconoclast who refused to remain sheltered in the ivory laboratory."

Mutinous in research and radical in policy, Commoner is an outsider both in academia and in government. Outspoken and iconoclastic, he is liked by the press and public, not by establishment scientists. And although he is the most self-conscious and clearcut example, he shares his maverick status with most of today's visible scientists.

All in the Family

In truth, the commandments derived from science . . . boil down to a diffuse, pervasive fear. . . .

— JACQUES BARZUN

On a cold January evening in 1973 at Stanford University, over two hundred people, the overflow from a packed auditorium, stood outside the giant lecture hall in a chilling rain to hear Dr. Linus Pauling's slow quavering voice piped over a loudspeaker. No one seemed to notice the irony of standing in the rain in order to hear Pauling's advice on how to prevent colds.

There is no question of Pauling's charisma or ability. "Pauling is arguably the greatest scientist alive today," science writer Graham Chedd has said. Pauling's contribution to the understanding of chemical bonds won him the Nobel Prize for chemistry in 1954. Some say his discovery of the cause of sickle cell anemia, part of his development of the concept of molecular disease, deserves another Nobel Prize. He has also made major contributions in the areas of antibody structure and function, protein structure, the relationship of abnormal enzymes to mental disease, and the theory of anesthesia. Pauling is not out of line when he explains that about "half a dozen fields of science were in some part, perhaps considerable part, based upon my original work." He is, by the estimates of friends and enemies, a genius. And he had the satisfaction of receiving the Nobel Peace Prize in 1963 for his political activities opposing war, particularly his work promoting a nuclear weapons test ban.

But Pauling's disarmament activities have won him as much

animosity as admiration. Because of his great stature in science, his public actions have received considerable attention. Both what he is saying and the fact that he, a scientist, is saying it, have received heavy criticism. The attacks sometimes depress him and his wife and four children, he says, and more profoundly, they have caused problems in his professional career.

In some respects, Pauling's troubles have been the obvious ones anyone faces when he ventures to publicize unpopular views. Pauling opposed nuclear weapons testing during the Cold War, urged peace and rapprochement with the Russians during the McCarthy era, and advised the public to take large doses of vitamin C at a time when megavitamins were anathema to the medical community. For his efforts he received the usual treatment afforded minority voices by the majority — ridicule, contempt, epithets, pressure to conform.

But in Pauling's case the problems were on a grander scale than for most, because of his position in the scientific community, the Nobel Prize, and the resulting visibility. What he said seemed to matter more and to receive more coverage. His scientific stature magnified the public's interest in his political views, and magnified its response. In 1943, at a time when Pearl Harbor was still fresh in people's minds and the Japanese on the West Coast were being herded into relocation camps, Pauling's wife hired a young man of Japanese ancestry as a gardener. The following night a Japanese flag and "some scurrilous remarks" were painted on the Paulings' garage and mailbox. "We received some threats through the mail, directed at my wife and children, and for two weeks the sheriff kept a twenty-four-hour guard around our house, while I was making a trip to Washington, D. C.," Pauling recalls. The gardener, an American citizen, had been inducted into the U. S. Army and had taken the gardening job while waiting for active service to begin. The incident was Pauling's first experience with political visibility.

Pauling's entanglement with press and public over the Japanese-American gardener foreshadowed many unpleasant predicaments after the war, as Pauling began to speak out against nuclear weapons. He came to his opposition of nuclear weapons almost inadvertently. Because he had worked on chemical explosives for the government during the war, he was asked by a local group, shortly after the atomic bombs were dropped in Japan, to give a speech on the differences between nuclear and more conventional explosives. Studying up for the lecture, he recalls in *The Humanist* magazine, he was appalled by what he learned about the ability of nuclear devices to destroy cheaply, easily, and perhaps inadvertently. He felt that people had to be alerted to the dangers, so that power-oriented governments would not be tempted to use nuclear weapons. He apparently prepared for the lecture with his usual thoroughness, because after the lecture, he recalls, an FBI agent visited him asking where he had gotten his facts. "I had to tell him the truth," Pauling told *The Humanist* interviewer, "that I had just figured it out on the basis of the available information about nuclear behavior." The same year he and Robert Oppenheimer called a meeting to form a Pasadena (California) Federation of Atomic Scientists which, like its better-known Chicago counterpart, worked to promote understanding of the dangers of atomic energy. In 1946, Pauling joined the Emergency Committee of Atomic Scientists (the "Einstein Committee"), chaired by Albert Einstein and dedicated to fostering education on atomic weapons.

Thereafter the press frequently found occasion to chronicle Pauling's political problems. From 1952 to 1955 a cluster of press appearances dealt with his irreverence for Senator Joseph McCarthy and McCarthyism, and his resulting problems. McCarthy and followers branded Pauling a communist, a label which he repeatedly denied but which still clings to him today. In spite of protest from the international scientific community, Pauling was denied a passport in 1952; later his passport was "limited," and finally restored only when he needed to go to Stockholm to receive the 1954 Nobel Prize for chemistry. There was another surge of publicity in 1957 and 1958 when Pauling initiated a petition among scientists urging a nuclear test ban treaty, and another in 1960–1961 when the Senate's Internal Security Subcommittee held hearings on the petition and Pauling refused to reveal the names of volunteers who had helped circulate it.

In each political episode, the press's coverage ranged from valiant defense to vicious attack. On the one hand, *The Nation*, for instance, drew an analogy between the Senate Security Subcommittee's harassment of Pauling in 1960 and Russia's treatment of novelist Boris Pasternak. On the other hand, in the same era, *U. S. News & World Report* took the occasion of Pauling's petition and hearings to trace the history of Pauling's connections with communist-affiliated organizations and to impute sinister motives to Pauling's efforts. The magazine concluded, "It is clear from his record and from his recent public statement in Chicago that Dr. Pauling believes that movements dedicated to peace and disarmament should collaborate with the Communists — or, as he put it, should be 'open to all.' "

Pauling's apparently ominous "open-to-all" policy prompted the editorially conservative *National Review* to single him out in 1962 as "acting as megaphone for Soviet policy by touting the World Peace Conference that the Communists have called for this summer in Moscow just as year after year since time immemorial he has given his name, energy, voice and pen to one after another Soviet-serving enterprise." He was, the magazine added later, "one of the nation's leading fellow-travelers."

When Pauling's Nobel Peace Prize was announced in 1963, an editorial in *Life* magazine branded the honor a "weird insult from Norway." Since Pauling was, in the Senate Internal Security Subcommittee's words, "the No. 1 scientific name in virtually every major activity of the Communist peace offensive in this country," *Life* said it was "an extraordinary insult to America" that the Nobel Peace Prize Committee conferred its prize on Pauling. The editorial, one of a number of similar attacks, seemed to mock the statement Pauling had made to the press when he learned he had received the prize; he said, "For many years it has not been respectable to work for peace. Perhaps the Norwegian Nobel Prize Committee's action will help to make it respectable."

In connection with the peace prize, Pauling also recalls that the chairman of the Norwegian Nobel Prize Committee was appalled by the lack of respect and enthusiasm America showed on the occasion. The chairman of the committee, who met Pauling at the airport, remarked that it was the first time in the history of the Nobel Peace Prize that the recipient had not been met at the airport by the ambassador of his country, also the first time the recipient's embassy had not given a reception. "He was really very upset," Pauling remembers.

In the meantime, Pauling had begun to sue for libel some of his more rash press critics, including the *National Review*. Pauling's case against *National Review* was one of the first in which the judge's ruling relied heavily on the 1964 decision in *New York Times* v. *Sullivan*, in which the U. S. Supreme Court held that a "public official" could not recover damages for libel unless he could prove "malice," a "reckless disregard" for truth. In Pauling's case, the New York State Supreme Court dismissed Pauling's charges on the grounds that, while Pauling was not a public official, he had become politically prominent as a result of his activities and public statements. Justice Samuel J. Silverman wrote, "Dr. Pauling has added the prestige of his reputation to aid the causes in which he believes. I merely hold that by so doing he also limited his legal remedies for any claimed libel of his reputation. And perhaps this can be deemed another sacrifice that he is making for the things he believes in." *Time* magazine, reporting the decision, called Pauling's defeat one of the "perils of being too public." *National Review*, both before and after the decision, was caustic and cocky, heralding its victory under the heading, "Linus Pauling — TKO."

There were other, more serious, perils of being too public, or too political, including effects on Pauling's scientific career. Ultimately, pressure from administration officials prompted Pauling to leave California Institute of Technology, in Pasadena, where he had taken his Ph.D. and taught for thirty-six years. His final decision to leave came in 1963, he says, because of "the ungracious reception of the peace prize by the Institute." He recalls that after the prize was announced, Cal Tech President Lee DuBridge made a remark "something like 'It's really a remarkable thing that someone should get a second prize, Professor Pauling; but there is of course a difference of opinion about the value of the work you have been doing.'" Pauling had been considering leaving Cal Tech for about two years, because he had been under continual pressure from the trustees and the administration to stop his peace activities.

James D. Watson writes in *The Double Helix* that Pauling had problems at Cal Tech as early as 1952: "Several members of Cal Tech's governing board . . . would have been delighted with his voluntary departure. Every time they picked up a newspaper and saw Pauling's name among the sponsors of the World Peace Conference, they seethed with rage, wishing there were a way to rid Southern California of his pernicious charm." As a means of applying pressure, Pauling says, he had been removed from his position as chairman of the division of chemistry and chemical engineering at Cal Tech, a post he had held for twenty-two years, in 1958, which was, he points out, the year his nuclear test ban petition was publicized. At the time the president of Cal Tech told Pauling that he would have been dismissed from the school altogether were it not for the fact that he had tenure as a professor. The trustees, who could not understand why the president could not simply fire Pauling, were furious, and one member of the board, Pauling was told, resigned in protest.

As a result of the 1958 episode, Pauling kept his professorship but at a reduced salary. More important to Pauling was the fact that his allocation of research space was reduced. Traditionally, a university gives its successful researchers, especially Nobel Prize winners, copious research facilities. Instead, in 1962, Pauling was again pressured to cut down on his laboratory space, this time by half. The suggestion was that this could be accomplished nicely if he would just eliminate his controversial work on the relationship of chemistry to disease. He finally agreed that he would do so beginning in the summer of 1964, and instead resigned in the fall of 1963. "Were it not for my political and social views," Pauling feels, "I'm sure Cal Tech would have found some way to support my research better and longer."

For the next three years Pauling worked at the Center for the Study of Democratic Institutions in Santa Barbara, California, where he pursued the theoretical aspects of his studies. He would have been "sunk," he says, if his work had not included theoretical aspects, because the center had no laboratory facilities. His laboratory research had to stop until he moved to the University of California at San Diego in 1967.

San Diego, however, was no more hospitable than Cal Tech. Pauling was already past retirement age (he was sixty-six), which meant his appointment was renewed on a year-to-year basis. He and Marxist philosopher Herbert Marcuse were the campus's two controversial celebrities, and the surrounding community voiced continual objections to their presence. When Pauling read an item in the newspaper in 1969, saying that Marcuse had been reappointed, implying that reappointments had been made and that he had been omitted, Pauling decided the sensible thing was to accept the job which had been offered at Stanford. Stanford University, Palo Alto, California, was also closer to Big Sur, he adds, which is where the Paulings are permanent residents.

Stanford, although it has a reputation for courting Nobel Prize winners, was not as hospitable to Pauling as might be expected. Pauling's office, in the new chemical engineering building, was cramped, and he had no room for a secretary until a janitor's closet was converted into an office. His laboratory was also unprepossessing and crowded, and in 1973 it was reclaimed for use by the chemical engineering department. There was no room in the chemistry building, where Pauling theoretically belonged, so, after disappointing negotiations with the university, Pauling found facilities in a modern office building being constructed a few miles from the Stanford campus, across the road from Stanford Linear Accelerator Center. He supports his new laboratory on his own research funds, and retains a tie with Stanford only as professor emeritus of chemistry.

Is he bitter? On the contrary: "The duty of the university primarily," he believes, "is to its younger people, building them up, and I am very pleased that when I was twenty-eight, twenty-nine, thirty years old, I got good support . . . from California Institute of Technology. . . . I was chairman of a division for twenty-two years; I know the problems of having old professors around and using up all the space." Furthermore, he points out, he was already past retirement age when he went to Stanford, and it is fairly standard procedure in such situations for a university to offer only a temporary appointment, in order to avoid having to make awkward judgments about whether an elderly professor is fit to continue. "Universities have to protect themselves from the old guys."

But doesn't he feel he is still productive? "My work continues," he agrees, "and I think in a sense that makes me exceptional." This is why, he explains, he feels justified in continuing to seek research support, to continue work which he thinks is going to be very significant. "It has originality — we're doing things no one else is doing, which is what I've always tried to do all my life."

While the inconveniences of moving from place to place occurred relatively late in Pauling's career, other disruptions may have damaged his earlier scientific work. In the early 1950s Pauling's politics may have made him unfairly the loser in the race to discover the structure of DNA, the "most golden of all molecules," the genetic material responsible for heredity. The importance of the DNA finding virtually assured its discoverers a Nobel Prize. In 1954, it was announced that James D. Watson, a twenty-six-year-old American, working at Cambridge University with Francis Crick, a British molecular biologist, had made the sensational discovery of DNA's double helix structure. Equally sensational, as far as the general public was concerned, was Watson's book, *The Double Helix*, a frank account of the bold, calculated, ungentlemanly race that produced the discovery. Among the book's revelations was the fact that the pair's closest competitor was Linus Pauling, and that Pauling's progress was stymied by his not having access to the photographs of DNA prepared by Maurice Wilkins and co-workers at Kings College, London. Pauling had requested prints of the photographs after reading about them in 1951, but Wilkins had written that he "wanted to look more closely at the data before releasing the pictures." Then, in May 1952, when Pauling was scheduled to attend a meeting in London, he began arrangements to visit Wilkins's laboratory, and would logically have been shown the pictures. A scientist who knew Wilkins "felt sure that if Pauling had made the trip, Wilkins would have shown him something." Instead, at the last minute at Idlewild Airport, Pauling's passport was removed by federal authorities. "Failure to contain Pauling," Watson remarks wryly about the motivations behind the passport removal, "might result in a London press conference with Linus expounding peaceful coexistence."

Does Pauling think he would have been first to find the double helix structure of DNA, had he seen Wilkins's pictures? He told a science writer in 1971, "I think I would; though missing the alpha-helix [structure of a protein] in 1937 shows that even rather simple ideas can be elusive"; he added that he "rather wished the [DNA] idea had been his." Does he feel his devotion to political problems hampered his ability to make the discovery? Pauling answered in interview with a question, "How can you answer . . . what you would do, if you had more time?" But, he reflects, he was having a great deal of political trouble at the time. In general, Pauling feels his political activities have indeed interfered with research. "I devoted half my time to these public affairs. If I had devoted myself wholeheartedly, if I had been able to devote all my time, to my scientific work, I'm sure that I would have made more discoveries."

There are also indications that scientific honors have been withheld from Pauling because of his political views. Beginning in about 1968, Pauling was apparently nominated every year for the National Medal of Science, the United States's highest scientific award. His name was consistently rejected by the White House in both the Johnson and Nixon administrations, presumably because of his stand against the Vietnam War, in particular, and his political positions in general. In 1968, Pauling was told by a member of the nominating committee that Johnson personally crossed his name off the list. *Science* magazine learned that during the Nixon years, "the White House refused at least twice to consider Pauling for the honor even though — as a Nobel laureate — his name was prominent on the slate of candidates drawn up by the official selection committee, which the President himself appoints." In 1975, the first time the Ford administration made the awards, Pauling's name, accompanied by a strong recommendation from the selection committee, was finally approved. *Conservative Digest*, predictably, called it "another a-Pauling Ford action."

Two characteristics in Pauling seem to lead him into his political and professional hassles. The first is his dedication to the cause of peace, a dedication which prompts him to spend costly hours in activities not related to his scientific work. In December 1947, while on a trip to England on board the *Queen Mary* with his wife and children, he wrote a pledge on the back of a large old cardboard sign that had announced one of his lectures. His pledge read, "In every lecture that I give from now on, every public lecture, I pledge to make some mention of the need for world peace." Pauling has spent large amounts of time on public affairs, in which he now includes his vitamin C discussion, "out of a feeling of duty, conscience, not because I liked doing it — also because my wife felt I was doing the right thing."

Pauling's wife, Ava Helen, has provided steady support for his public activities. In describing the "one scrap" he had had on television, when he had become very irritated with Lawrence Spivack on "Meet the Press," Pauling says that after the show was over, his wife got out of her chair and started after Spivack. "He could run faster than she could, so she didn't succeed in catching up with him. He got down the hall and out the door." Pauling and his wife met when she was a student in one of the classes he taught while working his way through Oregon State College; they were married the year he graduated from Oregon State, 1922.

Pauling does not take it lightly that his nonscientific activities have cost him a great deal of time. He economizes on time wherever he can. Students and reporters have noticed, for example, that an interview with Pauling lasts as long as he feels is warranted, and no more. There is a minimum of small talk, such as winding-down and leave-taking at the end of a conversation. When you talk with Pauling on a matter in chemistry, one graduate assistant observed, as soon as Pauling feels you have covered the ground, you may suddenly find yourself staring at each other for a couple of minutes, and you take your leave; his mind has taken off on a tangent he now pursues on his own. Or, as you are walking along chatting with him, he may suddenly stop, take out the pad and pencil he always carries with him, and jot down an idea. He never engages in chitchat on trivial topics, according to the assistant. "At dinner at his house, you might pick up and read an encyclopedia." The conversation would be, perhaps, about the animals in the ocean, "just intellectual stuff — never about how his aunt and uncle were feeling. His mind is too busy with other things." This is one of his problems getting along with university administrations, the assistant feels. "He is not a politician — I could never see him having tea with the dean, having tea with anyone."

When Pauling notes, therefore, as he did in a February 1973 interview, that he spent an hour and a half the previous day with a "man who is psychotic about the condition of the world, because I was so worried about him," he is noting that he has given away what for him is a precious commodity, his time. In general, he sees anyone who comes around, including a number of "crackpots — I waste some time on them, but I think it is sort of my duty to do it." Very often they want him to do something to save the world. If people call on the telephone instead, he asks them to write down their ideas in an article or letter; then, if he feels it is important enough, he will see them. The mail piles up, and he answers each letter, from anywhere in the world, conscientiously.

In addition to his sense of duty and dedication, a second factor that has exacted a price from Pauling is an idealistic stubbornness. He points this out himself in explaining how he came to tilt with McCarthyism. At the time of the McCarthy hearings, he says, it was clear to him that "there was an effort being made to suppress opposition to militarism. I reacted by being stubborn, just refusing to give in. I wasn't so interested in it as I was in my work, but I felt it was the wrong thing to suppress discussion."

As Pauling described in interview the events leading up to his publishing the controversial *Vitamin C and the Common Cold*, he saw an analogy between his irritation with McCarthyism and his irritation with members of the medical profession whose minds, he felt, were closed to the published findings of investigators on the benefits of vitamin C. Pauling's attention was drawn to the vitamin C question in 1966 by a biochemist who heard him speak in New York, when he accepted the Carl Neuberg Medal for contributions in internal medicine. In the speech, Pauling expressed a desire to live another fifteen or twenty years to see the advances medicine was likely to make in that time. The biochemist, Irwin Stone, later wrote Pauling that he would like Pauling to remain in good health for the next fifty years, and sent him his regimen for taking large levels of ascorbic acid, or vitamin C. Stone had been studying vitamin C for forty years and had become convinced of its salutary effects. Stone feels he "happened to hit him at the right time." The vitamin C idea, Pauling notes, tied in with his own concepts of orthomolecular medicine and psychiatry, including experiments investigating a relation between ascorbic acid and schizophrenia. Pauling and his wife began taking vitamin C in doses Stone recommended and "noticed an increased feeling of well-being," especially a striking decrease in the number and severity of colds. Pauling began to mention his impressions of vitamin C in lectures, including a talk he gave in 1969 at Mount Sinai Medical School. After he returned to Stanford, he recalls, he received a letter from Victor Herbert, "a distinguished physician, giving me the devil for saying anything about vitamin C." The vitamin C idea was just nonsense, the letter added, and Pauling was just aiding the vitamin quacks. The letter then asked if Pauling could name a single double-blind (controlled) experiment supporting the idea. Pauling wrote back that he could not, but that he had not read the literature and would check.

Checking the literature was not a difficult job, Pauling found; "I read the papers in the way that a scientist would read the papers: what is the evidence?" Others had read only the abstracts, Pauling says, and those not very carefully, ignoring the tables and the data. Pauling found the evidence strong. He wrote Herbert again, this time mentioning specifically an experiment by Ritzel, reporting that 1000 mg. of ascorbic acid a day caused a reduction of 61 percent in the number of days of illness from upper respiratory infections among a group of skiers. Pauling says that Herbert wrote back, "I am not impressed by Ritzel's paper." And Pauling answered, "I am not impressed by your saying you are not impressed by Ritzel's paper. It does not matter whether you are impressed or not; the important thing is, What are the facts? It is not justified for you just to laugh that off." Interpretation of the data, which was apparently the point of difference between Herbert and Pauling, was to be the most persistent cause of debate between Pauling and other critics.

By this time, Pauling was beginning to be irritated with Victor Herbert nearly the same way he had been with McCarthyism. Here was the evidence, yet doctors just rejected it. He decided the appropriate action was to "write a little book." Since he was not reporting original research, he felt it would not be appropriate to publish in a technical journal. Nor did he want to take the time for the research experiments that would be involved. "I was 69 years old," he reminded an interviewer for *The Humanist*, "too old to commit large amounts of time to setting up a research organization to perform an entirely new study; and there was not enough time to design a research program and carry it out. I'm at the time of life when I should be devoting my energies to stimulating the thoughts and research interests of younger scientists. I should be sharing the knowledge and experience I have accumulated. My role is properly that of scientific gadfly." Later, when he started writing the book, he did initiate theoretical ideas of his own, and these he published as technical papers.

Pauling wrote his "little book" in about a month, and rewrote it in another month, after comments from his publisher. Like most visible scientists, he seems to have a knack for writing at the popular level. A British nutritionist, Reginald Passmore, calls the book "first-class popular scientific writing." One of Pauling's colleagues has noticed that if he writes a paper, fully understanding the material, then shows it to Pauling, Pauling will rewrite it so that it is much clearer.

Attesting to the power of Pauling's visibility and credibility, *Vitamin C and the Common Cold* brought immediate attention to a vitamin C theory that other researchers had proposed, unnoticed, for years. In 1971, the year after the book appeared, sales of vitamin C doubled in California. Pauling notes that Hoffman-La Roche, the principal manufacturer of vitamin C, built a new plant, which produces twenty-two million pounds a year, "a capacity considerably greater than the total capacity of all vitamin C plants before the book came out." *Life* quoted a health food store owner as marvelling, "This stuff is Norman Vincent Peale and Christmas, it's Coué, yoga, Zen, pie-in-the-sky, and the pot of gold at the end of the rainbow — all rolled into one."

The book also drew prompt criticism, especially from the nutrition specialists in the medical profession. Nutritionists were quick to point out that Pauling was not a medical doctor, arguing that he was therefore speaking outside his area of expertise. They also noted that he had not done any experimental research on the subject himself. There was not enough evidence to evaluate the role of vitamin C, they added, and there were some experiments showing that vitamin C had no effect. There were also differences in how Pauling and his critics interpreted particular key studies; in the same experimental statistics, Pauling saw significant vitamin C effects, critics saw none. And the critics suggested that large amounts of vitamin C would be toxic, would have side effects, or would pass unmetabolized into the urine.

Journalists took the critics seriously, particularly Pauling's most outspoken opponent, Frederick J. Stare of the Harvard School of Public Health, who called Pauling "lost in the woods," stressed his lack of formal training in nutrition, and cited evidence that indicated large amounts of vitamin C were not metabolized by the human body. Before Pauling's book was published in 1970, *Mademoiselle* magazine juxtaposed Stare and Pauling and handed down a decision favoring Stare. The media, unable to sort out the evidence, tended to cover the vitamin C controversy humorously, stressing the faddishness and the incredible alacrity with which the general public was trying Pauling's advice: "The Vitamin C Mania," "The ABC's of Megavitamins." As if they sensed a weakness in the godlike scientist, some journalists made fun of Pauling for probably the first time. After an interview which both she and Pauling found uncomfortable and unpleasant, *Vogue* reporter Leticia Kent challenged Pauling's book in detail and described the tall, angular scientist as appearing "somehow avian — like a giant ostrich getting ready for a run." Pauling had forgotten the interview with Leticia Kent until he was asked about it. Then he recalled her as one of the very few reporters he had ever run across with whom he had difficulty. "I thought she was really just terrible. Will Rogers said, 'I never met a man I didn't like,' but he didn't say anything about women. I got in a sort of fight with her during the interview. She had not read my book, *Vitamin C and the Common Cold*, and was wasting my time by asking questions of me that she would not have had to ask if she had read the book."

In the meantime, Pauling was having his share of confrontations with professional journals, some of the best of which refused to publish his technical papers on vitamin C. Shortly after the book was published, Pauling submitted to *Science* a long article on the technical evidence linking vitamin C and the common cold. *Science* "fiddled around" nearly a year, according to Pauling, first asking him to rewrite and shorten the article, then turning it down. Pauling submitted the shortened version to the *Proceedings of the National Academy of Sciences*, which printed it, as it had an earlier article on the role of vitamin C in evolution. The next year, however, the *Proceedings* rejected a third Pauling paper, which extended the vitamin C discussion to its role in treating cancer. The Academy argued that the paper could raise false hopes in cancer victims. The chairman of the Academy's editorial board, John Edsall, was also quoted by *Science* as having "extreme mental reservations" about publishing Pauling's earlier papers on vitamin C, a charge Edsall later denied. The controversial cancer article was ultimately published in the journal, *Oncology*, after a delay of eight months.

Pauling's book also stimulated rumors that he was senile. A favorite topic at cocktail parties became speculation that Pauling was slipping. Spectators observing Pauling for the first time found ammunition for the senility theory in the fact that Pauling quite obviously has the high, hesitant voice, the gentle face, and the sparse white hair of the elderly. Those in personal contact with Pauling, however, knew his slow speech belied a quick mind, and promptly dismissed suggestions of senility. Arthur Robinson, Pauling's laboratory director, recalls a young San Diego professor at a party remarking of Pauling, "He may be senile, but how will we ever tell?" Pauling is so bright, Robinson explains, it would be beyond anyone else to tell he was losing ground. A graduate student admits that he had had doubts, but has always found Pauling right in the end. There have been times when he thought Pauling was wrong on a point, perhaps a computation; then two days later the student would come to the same figure himself.

Pauling was not without allies. Some journalists, particularly ones with prior experience of Pauling, were cautious in their assessment of the vitamin C controversy. Norman Cousins pointed out in *Saturday Review* that Pauling's theory should be considered in the light of the medical profession's tendency to resist new ideas, particularly on vitamins. "Resistance to vitamin therapy has become institutionalized — indeed, almost ritualized," Cousins wrote. He also pointed out that Pauling brings considerable knowledge and background to the study of vitamin C and that he has made substantial contributions to molecular medicine. He decides that the evidence is probably inconclusive at this time, but urges that Pauling "should be confronted only on the highest ground."

There were even some surprisingly friendly comments from within the medical profession. Shortly after publication of *Vitamin C and the Common Cold*, the magazine *Nutrition Today* published a series of reviews of the book, all of which were by doctors and all of which were highly critical. But in his editorial at the end of the series, Cortez F. Enloe, Jr., also an M.D., observed that Pauling's approach to the study of vitamin C, that of putting theory before proof, was an approach the medical profession would do well to copy, not condemn. He pointed out that Pauling, in the best tradition of chemistry and physics, had developed a theory on the basis of reason, leaving "no avenue of intellectual insight unexplored." In this tradition, Enloe continued, the theory is expected to suggest experimental directions which can prove or disprove the theory; Albert Einstein, for example, based his theory of relativity largely on reason alone. The medical profession, Enloe observed, has rejected Pauling's theory out of hand because it was unproved, instead of using it as an opportunity to explore new directions experimentally. "With our currently popular style of facing medical problems," Enloe says, "it is doubtful if Dr. Einstein would have ever got by a National Institutes of Health grants board."

Many of Pauling's critics within the medical profession were also, ironically, outside their area of specialty, as one doctor who reviewed Pauling's book admitted in an interview. The clinician added that he based his review on a superficial study of the book only, and wrote the article because the editor of the journal in which it appeared was a very good friend of his.

British writer Graham Chedd went to interview Pauling with serious reservations, but came away with a much more favorable impression. When he had first gone to see Pauling, Chedd had thought, "from what I'd heard and read here he was blundering into a field in which he himself had done no research; a field, moreover, unusually contaminated with uncritical quackery, and one in which the forces of true objectivity were apparently ranged against him." After the interview, however, Chedd concluded, "Even if he is wrong, my visit had convinced me it will prove only that elderly gods can have feet of clay, and not, mercifully, woolly heads. But I hope he's right; otherwise, my new jar of ascorbic acid BP will have been an awful waste of money."

Pauling's own best defender throughout the controversy, however, has been himself. At times he delves into personalities in his impatience with his critics, as when he suggested that Stare had misinterpreted a study by Ritzel "apparently because he couldn't read German well enough, or he was too lazy to read the tables." But in general Pauling stresses the data, and the need for more data, confining the debate to the question of how to interpret the experiments.

Perhaps the most disturbing "price" of the controversy has been that when Pauling considered conducting a large-scale experiment on vitamin C himself, says a close associate, he rejected the idea: he felt it would not make any difference, that it would be lost in the politics of the debate. Pauling later conducted a few small experiments on vitamin C, but failed to get funding. Turned down by the National Cancer Institute in the early 1970s, he says, "I was so discouraged I didn't do anything for over a year." In 1975 he submitted another application to N.C.I. and was again turned down.

Pauling has not been deterred from debating the vitamin C issue: he is firmly convinced he is right. And he feels this way at least partly because in the past he has usually been right. His scientific career has been successful from the beginning. When at the age of only thirty he won the coveted Langmuir Prize, he was, according to *Scientific American,* "hailed as a prodigy of American science." The magazine quotes Langmuir as calling him, prophetically, "a rising star, who may yet win the Nobel Prize." He had been a child genius; *Saturday Evening Post* relates that by the time Linus was nine, his father "was writing to the editors of newspapers for lists of appropriate books to satisfy the unbelievably precocious and voracious reading appetite of his . . . son." Pauling was awarded his Ph.D. from Cal Tech summa cum laude.

The vitamin C controversy follows a pattern which has often existed in Pauling's work: on the basis of existing information he proposes a new theory. The scientific establishment is at first skeptical, perhaps incredulous. Over time, however, experiments suggested by his idea and conducted by him or others support his concept and it achieves acceptance. This was the sequence in his work on protein structure, on antibodies, and on molecular disease and sickle cell anemia, as Pauling describes in a review of his work for *Daedalus.* More recently, his concept of orthomolecular psychiatry, which states that imbalances in specific substances such as ascorbic acid may be related to mental illness, has antagonized the medical profession. Typically, as in the vitamin C case, Pauling does not do the experiments himself. His role is to develop the theory, from which other scientists devise experiments to test the new scheme. As Thomas Kuhn, Bernard Barber, and others have described, new ideas are often resisted at first by the established scientific community. Pasteur ran afoul of the medical community, much as Pauling has. According to Barber, "Pasteur met with violent resistance from the medical men of his time when he advanced his germ theory. He regretted that he was not a medical specialist, for the medical men thought of him as a mere chemist poaching on their scientific preserves, not worthy of their attention." Albert Rosenfeld, *Saturday Review* science editor, finds the medical community disturbingly resistant to new facts and ideas, and calls it the " 'factifuging' syndrome."

"I have an open mind, I think," Pauling says. "Most of my contributions have involved the recognition of some misinterpretation of the existing information about the world and the development of some insight into the most profitable direction in which to try to proceed." Ironically, the controversiality of even Pauling's political views could also be considered partly a result of the fact that they are pacesetting. In the 1950s he advocated a nuclear test ban and detente with the Russians, which became established policy in the 1960s. Only time and further experimentation will determine whether, in the case of vitamin C, Pauling continues to be not just a gadfly but a leader.

"The Balance Sheet of Their Own Self-interest"

The price Pauling has paid, then, for his pacesetting scientific and political views has been high — personal insults, pressure from university administrations, distractions from career interests, governmental intimidation, professional scorn. To a certain extent, he has simply risked "the perils of being too public," the uncomfortable publicity which surrounds any controversial public figure. But he has had extra problems because he is a scientist. In having become a scientist, Pauling had joined a small, professional community, with its own special set of values and rules. Pauling's activities have conflicted with some of the basic principles of the scientific community, and his deviance has not been treated lightly.

A powerful system of social control operates within science to keep a tight rein on its members. As in any society, there exists among scientists a set of norms, what sociologist Robert K. Merton calls an "ethos of science," which protects and maintains the community's standards. In particular, the norms of science foster a climate that will produce good basic research. Qualities such as disinterestedness and objectivity, which have been found effective in making scientific discoveries, are cultivated in young scientists; and the value of basic research, the importance of devoting oneself to research over all other activities, is instilled. The model scientist is always the bench scientist, the one who devotes his full energies to the laboratory. Ironically, this model is maintained even though only a small proportion of the people and the money in science are actually in basic research.

The young scientist learns the rules as a student or trainee. Increasingly, the Ph.D. is a prerequisite to entering a scientific profession, and graduate school offers the ideal spot to train a beginning scientist in the values of his profession. There he is indoctrinated in the "cult of research and creativity," as Jacques Barzun puts it, and comes to believe nothing can be more important than research. "To suggest that practice, or teaching, or reflection might be preferred," says Barzun, "is blasphemy."

To protect its goals, science "punishes" scientists who do not toe the line. Scientists feel a sense of failure, ingrained from their training, for disobeying the commandments. And they feel the resentment, disapproval, and antipathy of colleagues. The pressure of fellow scientists on a transgressor is of crucial importance to him because the very success of a scientist in basic research depends on his ability to win the recognition of his colleagues. The recognition, the sense that his research has been evaluated and found competent and useful, is considered the scientist's highest reward, taking precedence over money, public fame, or other compensations. Reflecting the supreme importance of recognition, Albert Szent-Gyorgyi, a highly respected Nobel Prize winner, "once so despaired of any recognition in his early years in biochemistry," according to writer Mitchell Wilson, "that he actually determined on suicide."

The scientist is thrust into a highly competitive struggle for recognition, recognition in the form of citation of the scientist's work, awards, job offers, promotions, interpersonal approval, invitations to meetings, appointments to professional committees, and other honors. As the number of scientists increases, while the number of jobs decreases, competition becomes more and more intense, more personal, more frightening. By playing on scientists' fears and ambitions, the system of science provides efficient and effective quality control for the production of basic research.

There is, however, one hole in the system's network of norms: how are scientists to handle relationships with the rest of society? Without public goodwill, research freedom and funding are in jeopardy. Yet no clearly defined rules exist in science to cover this crucial part of the job of protecting and promoting research. Instead, the scientific community is a morass of conflicting and changing attitudes on the subject of communicating with society. On the one hand, it is fairly well accepted today that scientists must do some public relations work, some popularizing, in order to loosen the public purse strings. This view increases as funds decrease and fears of antiscience sentiment grow. Scientists, previously afraid they would be misunderstood if they *were* involved with popular communication, now find they are misunderstood because they are *not*. The situation, as science-watcher Spencer Klaw sees it, is "forcing scientists to become involved in the kind of politics in which all citizens must engage if they want large sums of money from the government."

On the other hand, because of the emphasis in science on the nobility and necessity of doing basic research, activities such as popularizing are viewed as a little lowly, distracting at best, demeaning at worst. Popularizing, administration, or other maintenance functions may be all right for the scientist who is past his prime, or who cannot make the grade in the laboratory, but they are not for the serious researcher. As a rule, scientists are also not well suited psychologically for the job of communicating with the public. Studies support the idea that many scientists are aloof, isolated from society, absorbed in their work, and uncomfortable in interpersonal and political situations. Besides, a scientist who speaks out, particularly in a political context, risks antagonizing his employer or his funding agency, even losing his job. John Gofman observes that his colleagues were "very frightened" at the time of his dispute with the Atomic Energy Commission. Gofman and Arthur Tamplin, as research associates at the AEC-financed Lawrence Livermore Laboratory of the University of California (then called the Livermore branch of the Lawrence Radiation Laboratory), spearheaded a bitter controversy by arguing that the AEC's radiation standards for nuclear power plants were not safe. Co-workers at the laboratory would tell Gofman, "Look, I agree with you, but I can't afford to agree with you. I have a wife and family to support, and I would be ruined." Gofman has since resigned from the laboratory, while continuing as professor of medical physics at Berkeley and heading his own company, Cardiodynamics, in Livermore, California. Tamplin has remained at Lawrence Livermore Laboratory.

In short, while certain moral, political, and social concerns encourage scientists' participation in public issues, other more pressing political, economic, social, and psychological needs militate against it. The more immediate needs for personal success usually outweigh the long-range needs for public goodwill and social good. As Philip Abelson, publisher of *Science* magazine, writes, the "balance sheet of their own self-interest" nets out in favor of silence. A newspaper interview with Nobel Prize-winning geneticist Arthur Kornberg epitomizes the conflict scientists feel about public communication. Kornberg had consented to see the reporter, he said, only because he was concerned about the crisis in confidence and consequent decreases in funds for basic research. It was not, Kornberg stated, his job to talk to newspapers; "it's not what I want to do or think I ought to do." And yet, in another context, he felt it was his job "to find the truth — and then not keep it from anyone."

The way the scientific community handles the conflict between long-range ambitions to inform the public and more pressing needs to concentrate on basic research is by making only particular, limited public activities acceptable. The scientist stays out of danger if he follows certain rules.

Rule one: he should confine his activities to the Washington advisory system if at all possible. The scientist who flies to Washington perhaps once a month to evaluate new research proposals or technological programs is part of a respected, selective elite, fulfilling scientists' public obligation in the approved way.

Rule two: the scientist should limit his public activities to a small percentage of his time. After all, research is his goal, and the rest is mostly distraction. Campaigning for candidates in political elections has become relatively popular, according to Stanford University sociologist Bernard Cohen, because it is seasonal, demanding a scientist's attention for only short periods at a time.

Rule three: the scientist should try to postpone most of his public efforts until after his most productive research years are over. Since it is also an adage of the scientific community that "science is a young man's game," an older scientist, especially a successful one, is allowed more time for public activities. Whether or not it is true, the adage becomes a self-fulfilling prophecy: during the life course of the typical scientist, the amount of time spent on research steadily decreases, and the amount of time on public activities, administrative duties, and "gatekeeping" functions (like refereeing scientific papers and distributing research funds) increases.

Rule four: the scientist should restrict his public communications to those that can be considered in his "area of expertise." He is only an expert on subjects related to his Ph.D., no matter how much he may have studied other areas on his own.

Rule five: the scientist should confine his remarks and activities to those that will enhance the public's image of science and its propensity to provide funding. Popularization is better than politics, because it is safer and extols the virtues of science. As a corollary, the scientist, by all means, should not dredge up and expose controversies that are raging in the scientific community behind closed doors. Ehrlich and Commoner's dispute, many felt, should have been a private one. Whether political or scientific (if there is a distinction), controversies will detract from the image of science and scientist as objective and rational.

Rule six: if a scientist feels he must express political opinions, he should keep them in the moderate range of the political spectrum, avoiding extremes. A liberal scientist speaking against the Indochina war to church groups and civic gatherings is more acceptable than a radical scientist making the same number of speeches to demonstrators and press. Physicist Charles Schwartz is "quite thoroughly alienated" from most of his colleagues, he says, and was dropped from the Lawrence Berkeley Laboratory staff in "retaliation" for his leadership in the free speech movement at the laboratory in 1970. Schwartz attributes his difficulties to the fact that his protests affect his colleagues directly. There is an "aura of liberalism" among scientists, he says, such that it is acceptable to protest on abstract issues, but his activities have hit "too close to home." In general, political activities that protect and enhance the scientific establishment are acceptable; those that threaten it, not surprisingly, are not. Since the scientific establishment is vitally connected to government and industry, activities that question the overall established structure of the nation are in the long run rejected.

The case of Joshua Lederberg demonstrates that a dedicated scientist's public activities can be very acceptable — if he follows the rules. Lederberg is a respected geneticist, chairman of the Department of Genetics at Stanford University School of Medicine, a winner of the Nobel Prize in medicine, and a member of the elite corps of "insider" scientists who serve frequently as advisors and consultants to government agencies. Yet, until recently (1966 to 1972), he also wrote a newspaper column, "Science and Man," for the Washington Post Syndicate.

His venture into the world of the popular press never met with the criticism he expected. On the contrary, he has had "very positive reinforcement," partly because colleagues feel "somebody has to do the job, and I'm getting it off their backs by taking it on for them." Second, he notes that he is doing it "at a level of rigor and respectability that meets their criteria."

For Lederberg, colleagues' standards happened to coincide with his own standards. He has always tried to keep his personality and personal life out of his popular communications, contrasting himself in this regard to most visible scientists. Rather than espouse a particular point of view or join one side of a controversy, he has tried to use the broad base of his expertise to bring important subjects to the public's attention. His mission, he says, has been to establish a more thoughtful, skeptical attitude in the reader, to elevate the public's level of interest in science and science's budget, and to improve the quality of science policy decision-making. One of his columns might describe a piece of basic research that appeared to be remote and abstract but turned out to have useful applications, or a development in genetic engineering that had important moral and political implications for the public to consider. Scientists could hardly fault his objectives or the reasoned, unemotional tone of his writing.

Lederberg also had the advantage of a virtually unshakable scientific reputation, assured by the Nobel Prize in 1958. By the time he began the column, he was at the stage in his career where nonresearch activities are increasingly acceptable, and he has continued to keep a hand in research. He keeps public involvement under control by limiting travel and by accepting about one percent of "what's thrust" at him. A quarter of his time, he estimates, is devoted to policy-oriented study and work. Lederberg even maintains credibility as an insider while continuing his outside role, by rigidly segregating the two. Confidences he receives as a trusted insider are not taken to the public; issues raised with government officials were not brought up in his columns.

Lederberg's decision to write the column was as cautious and reasoned as the columns themselves. He dates the beginnings of his serious thoughts about public communication to a symposium in London in 1962 on "The Biological Future of Man." The conference brought home to him the fact that the public and its government were not getting the information they would need to deal with coming scientific advancements in genetic engineering and other fields. His thoughts germinated for four years, culminating in his arrangement to write for the *Washington Post*. In an unpublished manuscript, Lederberg describes how he came to write for the *Post*:

In January 1960, returning from a meeting on space research in Nice, via London, I found that my seatmate was Nigel Calder, whom

I had already met very briefly as founder and editor of the English science news magazine, *The New Scientist*. The plane ride from Nice to London was a good occasion for us to discuss the problem of public information about science, for which I felt his magazine was doing a unique service. In October 1964, having exchanged a few casual thoughts in the meantime, he wrote what was to me a rather novel proposal, that I become a regular essayist for his magazine. This was particularly startling since my previous experience at popular writing had not been a very fruitful one, at least by editors' standards.

I had to say "no" to Calder's invitation but it did set me to thinking about the gap in communication between scientist and citizen and about the most appropriate format in which it would be possible for a scientist like myself to add a new kind of commentary about scientific advance. In other words, what kind of proposal would be so attractive that I would not refuse it, and then why not take the initiative myself?

After some thought I concluded that a regular, short column in a newspaper of wide, literate circulation could be the most effective channel that could be devised, at least for my own contribution to that gap.

During the next eighteen months I gradually put together some material for a prospectus for such a column and a few sample pieces. Fortunately, one of my associates knew some of the people involved in the management of the *Washington Post*, and helped to convey my material to them. In the course of time and with the particular interest of Mr. Howard Simons, who had just been elevated to Associate Managing Editor from having been a well-known science writer himself, the proposal for a weekly column was tentatively accepted and I have been enjoying this function ever since [to 1972].

Ultimately, Lederberg discontinued the column because he could not meet his own standards for public communication. Each week's writing raised a number of new issues, until he developed a bewildering backlog of unanswered questions. He decided, "I had done my bit. . . . I found myself becoming too preoccupied simultaneously with what ended up being over two hundred different issues that I had raised. . . . A lot of the material was off the top of my head, as good as anybody else's in this arena, but still didn't satisfy my own criteria for scholarship and depth." He did not have the reportorial temperament, he reflects, and could not do one article, then put it out of mind. Hardly a typical reporter in many ways, one might add, he remained very much a member of the scientific community.

Exhibiting once again the science community's sense that public activities should be limited, Lederberg concludes, "I think I've done more than my share in that regard." He does not expect to go back to a *Post*-type column again, but continues social involvement indirectly by advising other scientists interested in science-and-society issues.

Complicating the picture of scientists' attitudes toward public and politics is the fact that, in response to social pressures, the rules are changing. One physicist will maintain that politics is approved by the scientific community and should not be; another will protest that politics is not approved and should be. Sociologists find there appears to be no consensus at all on the question of whether scientists should consider the social implications of their research.

Science watchers are in almost complete agreement that scientists' sense of social responsibility is broadening. Both the number and caliber of scientists involved in public activities are increasing. University courses in science-and-society are infiltrating even hardened science departments, and the rhetoric of social responsibility is fashionable. "It's become established," Barry Commoner commented. "Sure, Phil Handler [Philip Handler, president of the National Academy of Sciences] will talk about social responsibility. As a matter of fact, it's sort of ironic, but what I and others like Margaret Mead tried desperately to get the AAAS to do, it's now doing in an establishment way."

Skeptics question whether the new professed sense of social responsibility runs very deep in the scientific community. The late Mitchell Wilson, a free-lance writer and physicist, felt that scientists' expression of concern for the outcome of their research is just talk. Most scientists, he said, want to do the work and get the scientific credit, but also want to take the new social-conscience stance. Others have pointed out that scientists often take a socially responsible stance to protect themselves from government interference or public hostility.

The durability of the new protestations of social responsibility depends on the durability of public antiscience feeling, job shortages, funding restrictions, and technological dilemmas. Since the nineteenth century in the United States, there have been periods when the acceptability of public activities increased and decreased, depending in part on the climate of public opinion. Historian Ronald C. Tobey notes that in the nineteenth century there was a tradition of popularizing science, accepted as part of contributing to the advancement of science; such leading scientists as Michael Faraday and Thomas H. Huxley frequently gave popular lectures. Scientists seemed to feel a need to justify their existence to the public, and to obtain social and financial support from the educated middle class. As science became more professionalized, accepted and supported by industry and universities, popularization declined, and was unacceptable in the scientific community by 1915. Following World War I, Tobey continues, a "new consciousness" among scientists involved in the war effort created a new interest in popularization; after World War II, the atomic scientists' movement and related activities represented a similar wave of interest. Perhaps the movement today to "politicize" science and engage in grass-roots activities, following a period in the 1950s when only insider, advisory activities were condoned, represents another wave of interest in public roles for scientists.

The visible scientists are testing the limits of scientists' protestations of social responsibility. Unlike Lederberg, most visible scientists today are breaking the rules, testing the rules' validity by a kind of civil disobedience. Speaking out publicly, prominently, politically, they are forcing scientists to examine how much they really believe in practicing the social involvement they preach.

What Price Success?

Altogether things have gone pretty well, so I don't complain — and I fight back.

— LINUS PAULING

"I expected it would totally destroy my scientific career," Paul Ehrlich said recently of his public visibility, "not because I expected to get out of research, but because the average scientist is basically toilet-trained to the point where if what he does is comprehensible to the general public, it means he's not a good scientist. That's what I thought. I was wrong. The reaction of my colleagues in population biology has been so close to one hundred percent favorable it's stunning to me."

Visible scientists circulate in science and society as if protected by an invisible shield. Criticisms abound, but careers do not crumble. To everyone's surprise, the ax never falls. As Carl Sagan says, "It hasn't at all been as negative as I had feared."

One form of protection for the visible scientist is academia's version of job security: tenure. Nearly all the visible scientists were tenured professors and scientifically successful before their popular activities began to demand much time or attention. Linus Pauling began protesting against war and the use of atomic weapons after World War II; he was a full professor at Cal Tech in 1931, although not a Nobel laureate until 1954. Ehrlich first became popularly known in 1969 and 1970; he was a full professor in 1966. William Shockley's first speech on dysgenics was in 1964; he joined the Stanford University faculty as a tenured full professor in 1963, after becoming a Nobel laureate in 1956. James D. Watson began writing *The Double Helix* in 1965; he was a full professor at Harvard by 1961, and received the Nobel Prize in 1962.

The sequence of tenure-then-trouble has not necessarily been deliberate. Several scientists began their public involvement before they were scientifically well-established, and achieved visibility later, perhaps partly because they needed a solid reputation in order to be singled out by the media. Nor have the visible scientists been entirely immune from job risks, as Pauling's experience shows. Ehrlich, as another example, did publish general articles on population in 1965, a year before he was promoted to full professor. At the time his promotion was considered, a fellow member of his department recalls, his popular articles were discussed: were they scholarly? It was decided that the articles were important anyway; as long as one carried on research of more than adequate quality and quantity, the other articles should not be taken into account.

Even better protection for visible scientists: money. Financially, a young teacher usually does better to concentrate on getting promoted; the increased salary will bring him more money than would the income from popular books or lectures. Paul Ehrlich, however, "would have to be promoted to God," one professor observed, to better his financial success from *The Population Bomb* and popular appearances. Feeling perhaps the independence which goes with being valued and valuable, Paul Ehrlich cavalierly filled out part of his annual faculty report to Stanford University, 1971–1972, as follows:

Q: Work completed during 1971–72:
A: Lots (see starred items on attached list of publications).
Q: Work in Progress:
A: Lots more.
Q: Other work planned or proposed:
A: Even more than that.

"I don't know how anybody could affect my career in a way that would really disturb me," says Ehrlich, "except by attempting to get my money from the National Science Foundation cut off; that would more hurt graduate students and post-docs than it would hurt me, because I can continue the things that are interesting to me with very, very little support. . . . I could almost certainly raise ten to twenty thousand dollars a year, which would give me minimal research support, on my own hook if I have to."

Does the scientific community put the squeeze on visible scientists' research funds? It is, after all, insider, establishment scientists who review research proposals in Washington. But the visible scientists have not found this a problem. "Research support has remained inadequate," Ehrlich says, "but it was inadequate before I got into this. And I guess everybody feels his research support is inadequate. . . . If anything, since most of the panels I go to for research support have the same concerns, I think they might lean over backwards in the other direction, if they could. I don't think there has been any leaning in either direction. I think they judge my research as they did before, on my research productivity, which is what they should do." Sagan agrees: "As far as I can tell, there have been no penalties for my visibility, and no benefits either. As far as I can tell, it's entirely fair."

Also insulating the visible scientists is the science community's own fledgling social ethic. Finding it unfashionable and uncomfortable to condemn scientists' public activities, critics vent their hostility indirectly. One approach is simply to divorce visible scientists conveniently from science: "They are no longer scientists," said Nobel laureate geneticist Arthur Kornberg concerning Ehrlich and Commoner; "they have become publicists or entrepreneurs." Margaret Mead finds, "What my dear colleagues do — and remember I am the only woman in the group — they meet me at meetings and say, 'Oh, my wife is so interested in your articles; she reads them at the hairdresser.' " Reflects Sagan, "The negative comments tend to be peripheral, told to someone who tells you. But the only people ever who tell me critical things are good friends of mine. The people who are upset virtually never say a word to me. . . . I've been nicely insulated from the hostile comments. So I've gone blithely on, not realizing there are people offended."

Like a political candidate or Hollywood star, a visible scientist is surrounded by a bevy of devoted followers, a protective coating of allies who fend off barbs and doubts. Pauling's closest associate, for example, is Arthur Robinson, his laboratory director, who headed Pauling's laboratory first when Pauling was at the University of California, San Diego. When Pauling moved to Stanford, Robinson took a year's leave of absence to set up Pauling's new laboratory, then commuted for two years, teaching at San Diego, running the laboratory at Stanford; finally, in January 1973, he moved to Stanford, resigning his post at San Diego to work full time for Pauling. A postdoctoral student who met Pauling at San Diego also changed his career plans and moved to Stanford with Pauling. He felt what Pauling was doing was worthwhile, whereas his own field, theoretical chemistry, didn't seem to be doing anyone any good. The people working in Pauling's laboratory are all staunch defenders of Pauling's controversial vitamin C theory. By contrast, Barry Commoner is one of few visible scientists to feel alienated from the scientific community, and he is a loner at Washington University, not well insulated by supporters.

The visible scientist's staff also safeguards his research when he forsakes his laboratory for the public forum. Commoner feels he can leave his laboratory for a whole month without problems. "I was lucky," Ehrlich says, "to have good people around so that things that were already started could continue on — and the credit belongs to them, not to me." In 1970, Ehrlich did fall behind in writing up his research findings, and had to catch up two years later. For Mead, the world is her laboratory: she observes her audiences while they watch her.

Colleagues of visible scientists often express the feeling that the visible scientist's research has decreased in quantity but not suffered in quality. A close collaborator of Ehrlich's pointed out that Ehrlich has continued to develop new techniques and theories. Although he would have spent more time in research, if it were not for his population concerns, he might well have been just "filling in the holes," which others could do as well.

Science fiction writer Isaac Asimov gave up his research at Boston University, but only because he found his laboratory work "undistinguished" — "respectable but insignificant." For his Ph.D. dissertation on enzyme reaction rates, Asimov says, "I turned out what may very well prove to be the most microscopic addition to the sum total of human knowledge." The dissertation's title is one of the few pieces of evidence available that Asimov is capable of writing obscurely: "The Kinetics of the Reaction Inactivation of Tyrosinase during Its Catalysis of the Aerobic Oxidation of Catechol." Holding up a thin binder containing a handful of technical articles, he said in interview, "I have got nearly two hundred volumes of my magazine stories and my paperbacks and all that sort of stuff. This is the sum total of all my scientific papers." Asimov had embarked on a scientific career on the assumption that he could not support himself as a writer — science fiction paid a half-cent a word when he was a student. Once his science fiction and popular science writing became lucrative, he made the break with academia cheerfully, albeit not calmly, as he describes in a squib for *The Magazine of Fantasy and Science Fiction:*

I kept having the impulse to quit my job and just devote myself to my writing, but how could I? I had spent too much of my life educating myself for this job to throw it away. So I dithered.

The dithering came to an end in 1957, by which time I had a new department head and the school [Boston University School of Medicine] had a new dean. The old ones had been tolerant of my eccentricities, even fond of them perhaps, but the new ones were not. In fact, they viewed my activities with keen disapproval.

What bothered them most was the status of my research. As long as I had written only science fiction, my research was not affected. My science fiction was written on my own time. No matter how hot the story, or how pressing the deadline, it was written evenings and weekends only.

Nonfiction was different. I considered my books on science for the public to be a scholarly activity, and I worked on them during school hours. I continued to carry a full teaching load, of course, but I gave up my research.

I was called on the carpet for this by the new administration, but I held my ground stubbornly and even a little fiercely. I said that I was paid primarily to be a teacher, that I fulfilled all my teaching duties, and that it was generally recognized that I was one of the best teachers in the school.

As for my research, I said, I didn't think I would ever be more than a merely adequate researcher, and that while my scientific work would be respectable enough, it would never shed luster on the school. My writing on the other hand (I said) was first-class, and it would bring a great deal of fame to the school. On that basis (I continued) not only would I not abandon my writing in favor of research as a matter of personal preference, but also out of a concern for the welfare of the school.

I did not manage to put that across. I was told, quite coldly, that the school could not afford to pay someone $6500 a year (that was my salary by then) in order to have him write science books.

So I said, contemptuously, "Then keep the darned money, and I won't teach for you."

"Good," I was told, "your appointment will be ended as of June 1958."

"No, it won't," I said. "Just the salary. The title I keep, for I've got tenure."

Well, what followed was a Homeric struggle that lasted for two years. Never mind the details, but I still have the title. Since June 1958, however, I haven't taught and I haven't collected a salary. I give one lecture a year and I fulfill some honorary duties (such as sitting on committees), but I am now a full-time writer and will continue to be one. And I am still Associate Professor of Biochemistry.

The criticism that does filter through the visible scientist's outer defenses runs a gamut of outrage, insolence, and scorn. Visible scientists are not nearly so oblivious to attacks as they like to appear. Commoner, who tends publicly to dismiss his unappreciative peers as "a bunch of jerks," was reluctant to appear in *Time*'s 1970 cover story for fear of criticism from colleagues, according to a former *Time* editor. (Ultimately he proceeded with the story.) B. F. Skinner is renowned for his contempt of critics. "I don't usually read my critics. . . . I didn't read the famous [Noam Chomsky] review of my *Verbal Behavior* until ten years later, when my students urged me to read it. I had read a few pages and saw he missed the point." (At the time Chomsky also did not have the reputation he had ten years later.) While

Skinner and his biographers paint him as unscathed by criticism, he nonetheless put together a scrapbook of reviews and reaction (eighty percent unfavorable, he estimates) to his book, *Beyond Freedom and Dignity*. He made another notebook by tearing and interleaving pages from *Beyond the Punitive Society*, a volume of essays that criticized *Beyond Freedom and Dignity*. Recalling Cornell philosopher Max Black's essay, Skinner says, "I counted twenty-three different names that he calls me in the article." Similarly, in 1970 a journalist observed that Skinner's copy of Joseph Wood Krutch's famous critique was "lined and triple-lined in pencil." And Skinner confesses to "some feelings of anger" when he is accused of being obfuscating and unclear. "I have written ten books and I have written them very carefully; if I haven't made my point, I don't plead guilty to lack of clarity. I try very hard to make the case. These people jump to conclusions about what I am saying and then attack it — it's very strange." In response to these feelings, in 1974 Skinner published *About Behaviorism*, an answer to his critics, a primer of behaviorism, "designed to straighten out people who have grossly misunderstood what it is all about."

Humanly sensitive, and prone to extremes, visible scientists either ignore criticism or overwhelm it with response. Says anthropologist Gregory Bateson of Margaret Mead's reaction to criticism, "She douches it with productivity." As an example, Mead writes in *Blackberry Winter* that she planned to return to New Guinea for another field trip in the spring of 1931. Shortly before she was scheduled to leave, there appeared a review by Bronislaw Malinowski of her book *Growing Up in New Guinea* accusing her of not understanding the tribe's kinship system. "I was so enraged that I got our next field trip postponed for three months while I wrote my monograph *Kinship in the Admiralty Islands* to demonstrate the full extent of my knowledge of the subject."

At the other extreme, visible scientists, an egotistical group by all reports, dismiss all critics with equal disdain. B. F. Skinner was "forcing" himself, he revealed in a 1973 interview, to read ten pages a day of his notebook of critical essays from *Beyond the Punitive Society*. But, he added, "I do not see any valid criticism of my position." Skinner recognized that by ignoring criticism he left himself little opportunity to change his views. "I haven't changed much," he reflected, "and of course this makes you wonder whether you aren't stubborn or holding to some monolithic position that is vulnerable to criticism. No. I think I happened to hit on something that has paid off very well and I have stayed with it."

"Oh, sure, I get criticized all the time," says Ehrlich, "but I think of Aurelio Peccei, the Italian industrialist, saying of standard economists: 'they just make fools of themselves.' I don't even feel much of an urge to respond." Ehrlich and coauthor Richard Holm answered a letter to the editor that criticized their 1962 *Science* article with a one-sentence rebuttal: "Amadon's approach to problems of population biology emphasizes the need for unemotional evaluations of the concepts in this field." Taken to task for their "flippant reply" by a subsequent letter, the authors later made a longer comment.

In general, however, the visible scientists may be listening more than it appears. In 1971, Paul Ehrlich published a revised edition of *The Population Bomb* in which several passages that had been criticized were toned down. Critics had complained that he emphasized population control to the exclusion of other environmental problems; in the new edition, the "Prologue" continued to emphasize the fundamental nature of population in today's crises but deleted the last sentence of the earlier prologue, "Population control is the only answer." Third World critics had called him a racist for his frank description at the beginning of Chapter I of the emotional impact of driving through a Delhi slum. The old edition concluded:

Old India hands will laugh at our reaction. We were just some overprivileged tourists, unaccustomed to the sights and sounds of India. Perhaps, but since that night I've known the *feel* of overpopulation.

The new edition emphasized sympathy rather than revulsion:

Old India hands will laugh at our reaction. We were just some overprivileged tourists, unaccustomed to the sights and sounds of India.

Perhaps, but the problems of Delhi and Calcutta are our problems too. Americans have helped to create them; we help to prevent their solution. We must all learn to identify with the plight of our less fortunate fellows on Spaceship Earth if we are to help both them and ourselves to survive.

Ehrlich also deleted from the new edition a policy suggestion that, following a plan outlined by Paddock and Paddock in 1967, food should no longer be distributed to countries like India where there is no hope of avoiding the coming famine and death.

Many rewards besides the obvious financial ones weigh against the risks in the visible scientist's "balance sheet of self-interest." Nearly all of today's visible scientists like the limelight, even when they complain of lack of privacy and hectic schedules. Once a person gets a taste of it, fame is addictive, Gregory Bateson believes. New to stardom, Paul Ehrlich recalls that after his first appearance on the Johnny Carson Show, "you're sitting on an airliner and everybody comes by and sits next to you for a few minutes and stares at you — it's just absolutely mind-boggling nonsense." One senses that visible scientists enjoy their fast and full life style, even the parts they complain about. René Dubos, with characteristic self-honesty, said, "Writing is very hard work, but it's a kind of addiction." There is an underlying satisfaction in the visible scientists that they at least have put ideals into action. By taking an unconventional course, they feel fairly sure they have done what they, and not a peer group, wanted.

B. F. Skinner: High Risk, High Gain

The scientist who best reflects the curious mixture of frustration and satisfaction that accompanies popular success is B. F. Skinner. That Harvard University professor Burrhus (his mother's maiden name) Frederic (Fred to friends) Skinner is successful is undebatable: by psychologists and press alike, he is considered one of the most important figures in psychology. According to *The New York Times Magazine*, "in a survey of department chairmen at American universities . . . Skinner was chosen overwhelmingly as the most influential figure in modern psychology." From *Psychology Today:* "For perhaps the first time in American history, a professor of psychology has acquired the celebrity of a movie or TV star."

In the process of becoming the most visible psychologist, however, Skinner has had very bad moments. In particular, he has been roughly handled by the press. In its persistent image-making, the press has cast Skinner in the role of the mechanical man, unfeeling, routinized, complacent, cocky, as cold and methodical as his theories. Says biophysicist and Skinner-supporter John Platt, "Skinner may have had the worst press of any great scientist since Darwin."

To many laymen, Skinner's theories are indeed cold and dehumanizing. The underlying message is that people's behavior is determined by a combination of genes and environment, not, as they prefer to think, by free will, emotions, attitudes, impulses. Just as pigeons and rats can be taught to play Ping-Pong or run a maze by the proper schedule of reinforcements (operant conditioning) in a "Skinner box," people can be manipulated by shaping their environment. To Skinner, however, this state of affairs is a hopeful one, because by using the appropriate "behavioral technology," man's environment — and therefore his behavior — can be improved. In his most widely read book, *Beyond Freedom and Dignity*, Skinner contends that, with population explosion, famine, nuclear weapons, pollution, and other evils threatening man's existence, we can no longer afford the luxury of thinking of ourselves as free, dignified individuals; a technology of behavior is a necessity.

Critics have been quick to point out the chilling consequences of Skinner's theories. If man's behavior is to be controlled not by the individual, but by the scientists who have the technology to manipulate the individual's environment, the system is authoritarian, to say the least, and "fascist," in the minds of some opponents. To psychiatrist Thomas S. Szasz, Skinner is "just another megalomaniacal destroyer, or would-be destroyer, of mankind."

Others are scornful of Skinner's leaps of inference from rats and pigeons to people, what philosopher Arthur Koestler calls the "ratomorphic fallacy," attributing to human beings only the mental processes he has found in lower animals. When it comes to Skinner's particular vision of the new society, set forth in *Walden Two*, tastes differ. Enthusiasts have set up Walden-Two communes with mixed success. Critics led by Joseph Wood Krutch have denounced the Walden-Two idea as a threat to civilized life itself. Skinner's students, less involved in the heady issues of free will and determinism, have reflected that the Walden-Two life style sounds bland and boring.

Journalists, confronting Skinner's theories, have tended to catch the same chill that infects his critics, and the feeling has spilled over into their descriptions of him as a person. Profiles of Skinner evoke images of a man as well programmed as his laboratory and with as much compassion as a rat. They invariably mention his rigorous schedule, for example. According to descriptions of his graduate school days at Harvard, Skinner "rose at 6 A.M., studied until breakfast, went to classes, studied until 9 P.M., and went to bed. For two years he saw no movies or plays, had no dates [Skinner says 'scarcely any'] and read nothing but psychology and physiology." To improve his efficiency today, according to another account, Skinner keeps a cumulative record of the serious time he spends at his desk, just as he used to keep a cumulative record of a rat's performance in the "Skinner box." He has rigged a clock to start running whenever he turns his desk light on, and plots his work-time on a graph. *The New York Times Magazine* quotes Skinner as saying, "I treat myself exactly the way I treat my rats."

Journalists also concentrate on the cold-hearted aspect of Skinner's experiments. A favorite example is a mental hospital applying "reinforcement therapy inspired by Skinner," described by *Time* as follows:

The staffers of one institution . . . were troubled by patients who insisted on trailing into the dining-room long after the dinner bell sounded. Attendants tried closing the doors twenty minutes after the bell rang, refusing admittance to those who showed up any later. Gradually, the interval between bell and door closing was shortened to only five minutes, and most patients were arriving promptly. "You shift from one kind of reinforcement — annoying the guards and getting attention — to another, eating when you're hungry," says Skinner. To charges that this kind of conditioning is sadism, he replies that "the patients are going in quickly because they want to." That is strange logic: he seems to ignore the fact that the patients are compelled to "want to" unless they care to go hungry.

Of Skinner's pigeon experiments in the 1940s, *Time* says, "Dr. Skinner also exploited pigeons. By subtle and gradual application of the 'stretch-out,' he forced one pigeon to peck 35,000 times in a five-hour period for only one-third of an ounce of feed."

Another part of Skinner's media image shows him to be a thorough-going egotist. A 1968 profile by Berkeley Rice in *The New York Times Magazine* depicts him, as Skinner puts it, as a "conceited ass." The article, entitled "Skinner Agrees He Is the Most Important Influence in Psychology," emphasizes his confident attitude toward critics and comments wryly, "No one has ever accused Skinner of excessive humility." Skinner feels that the Rice article "seriously misrepresented me as I think I am." Skinner maintains that the author had to stretch the truth to make his point. The writer attended one of Skinner's classes, for example, in which Skinner says he discussed ways of working on one's verbal behavior. "Every one of you has something to say," he remembers telling the class, "and you really ought to take seriously the problem of getting it out. Every one of you is absolutely unique . . . you have an absolutely unique history as a person and this makes you unique — a unique moment in history." Instead, according to Skinner, "what this chap did was to say, 'Skinner regards *himself* as a unique moment in history.' Now that is just plain dirty dealing, because what I said was that *everybody* was a unique moment in history. Now that's in the literature. People will always dig that up. The obituary writer for the *Times* has probably got it in my obit, which is probably already written. And I'm stuck with it. . . .

"Well, it really hurt. I got that article in Paris. My wife and I were attending a UNESCO function there and then we [vacationed] for a month. . . . Every day I had this sinking feeling, 'My god, how can I go back and face my colleagues?' " When he returned from his trip, Skinner says, the editors responsible for the article apologized. Two years later another *New York Times Magazine* profile gave Skinner's telling rebuttal: "It would be absolutely inconsistent of me to take credit for anything I have done. The 'credit' goes to my genetic and environmental history." All told, Skinner is, understandably, "not too happy" with the way he has been treated in the press, an attitude that sets him apart from most visible scientists.

The journalistic image of Skinner comes partly from a professional image which is equally unflattering. Journalists who consult Skinner's colleagues find almost unanimous agreement that he is abrasive, difficult, impatient — "it would be almost impossible not to have something of a one-sided view," a sympathetic colleague admits. Close friends explain that the problem goes back to graduate school. As a graduate student, Skinner did most of his research himself, rather than in collaboration. "He was not the kind," psychologist Fred Keller recalls, "to discuss his hopes or plans or half-analyzed data around the laboratory coffee-pot, at the dinner table, or with a drink in hand at some convention. Such prudence may win respect, but is unlikely to initiate joint enterprise." Skinner has always been a loner, friends say. "He had an eye for the future . . . and was thinking of his place in the scheme of things," one colleague reflected. In addition, friends recall that Skinner's research was unusual, closer in many cases to physiology than psychology, so that he rarely referred to other psychologists.

His research was not only unusual, it was also highly successful, and resentments inevitably developed: "He was too damned successful for his age." Being highly successful, he did not suffer the humiliating setbacks common to most people; "he didn't get tamed the way most of us did," according to a friend. Over the years, while he "softened up somewhat," his reputation for abrasiveness grew. Also, while the professional image influenced journalists, the media image in turn undoubtedly influenced psychologists.

Skinner's public image is magnetized by the mechanistic in his

life and work. Skinner himself is very human. Elderly, pro-
fessorial, reserved, and intensely organized, he is nonetheless
friendly and trusting with interviewers, generous with his time,
enthusiastic about his family, honest about his failings, worried
about his health. But he also loves to invent. Growing up in
Susquehanna, a small town in northeastern Pennsylvania, Skinner built wagons, rafts, merry-go-rounds, water pistols, even a
steam cannon made from a discarded boiler, with which he
could shoot plugs of potato and carrot over his neighbors' roofs.
And thus he poses an image-maker's dilemma: his motivations
are humanistic, his solutions mechanistic. An engineer at heart,
Skinner translates concern for people into technological fixes.
"Even 'Walden Two,'" Richard Herrnstein has observed, "is
just engineering — human engineering."

Humanistic motives are not titillating to press and public, but
human technology is. Skinner's first taste of visibility was for
his invention of a mechanical baby-tender, and the reaction was
predictable. He had designed the "air crib" when he and his
wife decided to have a second child, to help cut down on the
disheartening number of chores connected with caring for an
infant. In place of a crib, their new daughter was kept in an
enclosed compartment in which temperature and humidity
were controlled, so that the baby would be comfortable without
clothing, and in which air was filtered, so that the baby stayed
cleaner and freer of disease.

In 1945, Skinner sent an enthusiastic article to the *Ladies'
Home Journal*, describing the success of the air crib in caring
for their new daughter, Deborah, then eleven months old. Deborah was happy, healthy, cheerful, developing well; she cried
rarely. Many of the problems caused by clothing, such as rashes,
danger of suffocation, and obstruction of movement, were eliminated. Mother was saved many laundry and bath chores, and the
stresses of tending a cranky infant. With alacrity and enthusiasm
that is rare in considering an unsolicited article, the editors of
Ladies' Home Journal sent photographers to Indiana and printed
an elaborate story on the "baby in the box," dominated by Skinner's own glowing account.

Deborah became "the most talked-about infant in America,"

but unfortunately for Skinner and his daughter, the talk was mostly negative — "there was hell to pay." People envisioned a baby trapped like a rat in a Skinner box, and were surprised when he protested that the baby was taken out of the air crib for diaper changes and play times. People thought, Skinner says, "you lock the baby up and look at him a year later to see what you've got." As Deborah grew older, rumors circulated, particularly among clinical psychologists, that she was psychologically damaged, if not psychotic. An English critic who traveled to the States one summer, Skinner recalls, said to a friend, "Isn't it too bad about Skinner's daughter, the one that was raised in that box, killing herself?" The friend answered, "Oh, when did she do that? I was swimming with her yesterday." The closest Deborah has come to psychosis or suicide, she says, was a typical identity crisis for six months after college. She is an artist, and her father proudly displays her modern paintings in his office and home. After her 1973 marriage in London, Skinner described his fatherly reaction: "They have been living together for a couple of years; they have a very nice flat in Highgate, and it is one of those very good things. A year and a half ago we were over there for Christmas, and his father and mother were there. We accepted [the living situation] perfectly. Then they came over last summer and got out a bottle of champagne and announced they were going to get married. I am a permissive parent: 'go ahead, if you want to.' "

Deborah feels her two-and-a-half years in the air crib were beneficial. The only disadvantage was the subsequent rumors, which she says her father feels may have affected her self-confidence. He explains, "I think at one time Debs was rather bothered by this — you know, her friends would take a course in clinical psychology and be told that she had become psychotic. I don't think one likes to feel that there are stories of that kind going around. Now she laughs it off, of course, as we all do."

Skinner's older daughter, Julie, has used air cribs for her two children, and Skinner is already considering how to find one of the scarce cribs for Deborah's future offspring. An Air Crib Corporation ventured to manufacture Skinner's invention but the owner died, Skinner says, and his son said he did not have the finances to continue. Devotees buy used air cribs for nearly the price new ones once cost, and thousands of people, according to Skinner, build their own. "My experience with American industry has been very sad," says Skinner. "Nobody ever took up the air crib properly."

Interest in his family also sparked another of Skinner's inventions, one that has been adopted considerably more widely: the teaching machine. Visiting his daughter's fourth grade class on Father's Day, Skinner was appalled by the way mathematics was being taught. The incident prompted him to develop the teaching machine and the principle of programmed instruction, now used extensively in American classrooms.

Intent on improving his own schedule — by engineering his environment — Skinner recently designed a gadget in his basement at home for getting exercise while also reading a book. Skinner rigged a belt on rollers, based on the principle of the treadmill, which allows him to walk in place, added a reading stand, and arranged a light over the stand. A timer is also attached, which he sets for twenty minutes. One of the books likely to be on the stand is Fowler's *Modern English Usage*, which Skinner rereads every two or three years.

Skinner had started out to be a creative writer. At Hamilton College he majored in English and wrote poetry and short stories. The summer before his senior year he attended the Middlebury School of English in Vermont, where one of the teachers invited him to have lunch with Robert Frost. Frost asked to see Skinner's work, and Skinner sent him three short stories. The following April Frost responded in a letter which concluded, "You're worth twice anyone else I have seen in prose this year." Frost's encouraging response clinched Skinner's decision to become a writer. When he tried writing in an attic study at home in Pennsylvania, however, "the results were disastrous." He took a "hack job" writing for coal companies, then spent six months of "Bohemian living" in Greenwich Village. Today he feels that the problem was that he had nothing important to say, but at the time he blamed literature and writers, who did not understand human behavior. "I was to remain interested in human behavior," he says, "but the literary method had failed me; I would turn to the scientific. . . . The relevant science appeared to be psychology, though I had only the vaguest idea of what that meant." Two years after embarking on a writing career, he started graduate school at Harvard in psychology.

Personal and humanitarian concerns were also in the back of Skinner's mind when during the summer of 1945 he wrote *Walden Two*. In the fall he was scheduled to move from the University of Minnesota, where he had taught for nine years, to Indiana University, where he had won an appointment as full professor and chairman of the department of psychology. "I rather suspect that it was the insecurity of that move which led me to write *Walden Two*," Skinner reflects. "There is a story back of how I came to write it which doesn't exactly jibe with that, but then there is more than one cause back of these things."

The story usually told in biographical sketches is that a woman at a dinner party in Minneapolis prompted Skinner to write *Walden Two*. At the party Skinner lamented that veterans returning from the war would face the same old problems they had left; a fellow dinner guest challenged him to write down his ideas for a better life. "To my surprise," Skinner is quoted as saying, "I began to write 'Walden Two,' " And he finished it, having written some of the novel in "white heat," seven weeks later.

The incentive was stronger than one dinner-party remark, however, "I think at the time I was dreaming of a better world," Skinner says. "I wasn't at all sure I was going to find it in Indiana. I was becoming an administrator — I didn't want that — and I got out of it as soon as I possibly could. Moreover, the campus wasn't very interesting there, and my wife hated to move. She had friends in Minneapolis, and never thought she could make friends [in Indiana]. That wasn't true as it turned out, but she had that feeling, and was in tears much of the time we were planning to move there."

Also at that time, just after his second daughter was born, Skinner says he was bothered by what happens when an intelligent woman like his wife has a couple of children, and "just deteriorates and becomes a housewife. . . . I really wrote *Walden Two* for the sake of feminine liberation, but very few women liked it. . . . If you tell an ordinary woman, American mother and wife, that she doesn't need to cook, if she doesn't want to, and doesn't need to clean house, if she doesn't want to, then her reaction is, 'Why would anybody love me?' It is a terrible scandal! It's awful." Later he told Germaine Greer he thought *Walden Two* would attract women's liberationists, and she asked him to send a copy. "I sent it to her and I never heard from her."

According to a colleague who was a student in Skinner's laboratory in the 1950s, a wealthy Asian came to the laboratory after *Walden Two* was published, proposing to set up a Walden-Two commune; Skinner spent a lot of time with him and considered joining, but his wife was against it. Skinner says he never actually came close to joining a Walden-Two community.

At first, Skinner had trouble interesting a publisher in *Walden Two*, which, although finished in 1945, was not published until 1948. He sent it first to Appleton-Century-Crofts, which had published his first book, *The Behavior of Organisms*, seven years earlier. "They thought it was talky-talky; they didn't see the point; a good many people didn't." Next, there was a traveling man from Houghton Mifflin, "a very interesting chap who used to come in and talk — he was a Unitarian, based in Indianapolis, and he tried to get Houghton Mifflin to take it. . . . He loved it, but they didn't want it." Later, after it was published, Skinner saw him manning a booth for Houghton Mifflin at an American Psychological Association meeting, and sneaking over to the Macmillan booth to read *Walden Two*.

Macmillan took it, but on the condition that Skinner also write a psychology textbook for them.

"They expected a text of the [usual] variety, with pictures, graphs, and so on. What I did instead was *Science and Human Behavior*, which doesn't have a picture in it of any kind, or a graph, or any data, and I am sure they were bitterly disappointed with that."

Macmillan published *Walden Two* expecting to lose money,

and "they couldn't believe their luck when it began to attract attention." Skinner thinks an editorial in *Life* attacking the book helped, as did Joseph Wood Krutch's critique in *The Measure of Man*. Surprisingly, Skinner was not criticized by his colleagues for writing a novel. Says Skinner, "Harvard professors have done all sorts of crazy things." Also, a colleague points out, *The Behavior of Organisms* was published first (in 1938), and it was highly theoretical and highly respected. "It was not as if he hadn't shown us his tables first."

Like most of Skinner's books, *Walden Two* was slow to catch on. At first, he found, nobody paid any attention to his new novel. "Neither did I. I didn't use it as a textbook until the middle '50s. I don't know why: I was a little embarrassed, I guess, to be using two of my own books." During the first twelve years after it was published, *Walden Two* sold less than ten thousand copies; in the next twelve years it sold over a million. At a million and a half, the plates wore out and in 1976 it had a new printing with a new preface. Skinner plots graphs of the sales of his books, and finds the pattern for *Walden Two* typical of all his books. After a kind of gestation period, there is a delayed response. "I had thought of making a mural of the growth curves of my books, because every single one of them shows something like that: there is a burst at the first, then it goes off for about ten years, and then it starts up. *Verbal Behavior* is starting up now, and people are beginning to get courses using it. The psycho-linguists of course never understood it and aren't even in the same ballpark, but there are those who begin to see what I was saying. I think it is my most important book, really, in spite of the criticism it has received." Even his first book, *The Behavior of Organisms*, was ignored at first, Skinner feels. "Nobody gave it a tumble for ten years . . . ten years, but I never once doubted the importance of that book." It is now a classic among behaviorists.

Skinner says he does not expect many people to be aware of his work until it has had time to grow naturally, and he is content to wait. "I am convinced that I am right on these things," he says, "and that eventually that will be seen — and that has proved to be the case."

Partly because of the pattern of sales from previous books, Skinner was surprised by the early sales of *Beyond Freedom and Dignity*, of which a million copies were in print in 1973. "We thought possibly thirty thousand as a maximum. I made a trust for my two daughters and gave it to them before it was published. Of course, now the money is pouring in. . . ." Skinner thinks sales were stimulated by the book's prepublication in *Psychology Today* and by a cover story in *Time*. He also made a number of appearances promoting the book, at the publisher's request. "They said, 'Would you go on television?' I said, 'Okay,' and the next thing I knew I had a two-page schedule of television and radio shows — I did forty altogether."

A reviewer for the *Saturday Review* called the stir over *Beyond Freedom and Dignity* a "delayed reaction," since the book was basically "a nonfiction statement of the principles behind *Walden Two*." Says a more acid critic in *The Atlantic*, "He states his old position more simply and therefore more ruthlessly than ever before." The book was attacked by everyone from Noam Chomsky, with whom Skinner has been having an "intellectual prize fight" for years, to then Vice President Spiro Agnew, who was quoted in news reports as calling Skinner "a serious threat to liberty." A magazine called *Freedom*, published by the Church of Scientology, featured a grotesque cartoon of Skinner's head on a rat's body, nose pressing a button. At one point a congressman began preparations to investigate the fact that the book was produced while Skinner was supported by a ten-year $283,000 Research Career Award from the National Institute of Mental Health. The congressman was later jailed for income tax evasion, Skinner says, and the investigation was dropped. "The attention lavished on Harvard psychologist B. F. Skinner and his new book is nothing short of remarkable," announced the cover of *The New York Times Book Review* on October 24, 1971.

In spite of believing that he will ultimately be proven right, Skinner says, "I start every morning at my desk as if I was just beginning. I don't take anything for granted." The first volume of his autobiography, *Particulars of My Life*, was published in the spring of 1976, and he is at work on the second volume. Volume

II, covering graduate school to the present, is "more difficult" for Skinner, because it involves collating "a fantastic amount of material," and introducing the general reader to his technical work. Again, he doesn't plan to generalize: "it really is particulars." Generalities will be in a third volume, "a behavioral interpretation of my life as a behaviorist." Sometime there might also be a short book on his techniques for self-management, ways to "get yourself to write, to think, to turn out what you are capable of turning out." Even later, he might try another novel. "I have thought about it a lot . . . it should be more or less autobiographical, because the theme would be the one I have most intimately experienced, namely, the position in the world today of the behavioral scientist, and the necessary conflicts he encounters. This is the theme of the last third of this century, and with a little luck I could write the novel that best expresses it." A "book about the future" has been shelved temporarily, after going through two drafts in 1975, because Skinner is "thoroughly sick of it." A new collection of essays can be expected someday, and a "psychologist's notebook," a collection of his notes on life, literature, science — "interesting little things — psychologists will steal lecture material out of them for years." His editor, he says, thinks the notes should be published posthumously.

Skinner's failing eyesight figures heavily in his plans. He has already lost some sight from glaucoma in the right eye, and he has a distorted retina in the left. The condition has "stabilized," but he no longer reads for pleasure. He figures he could finish his autobiography even if he lost his vision, although he does consult many documents, including notes from his mother and grandmother, as well as his own correspondence. "I find that my memories are not accurate, when I go back and look. . . . I was convinced until fairly recently that in the middle '30s I had only one offer of a job, which was at Minnesota, which I took — with the exception of one that I wouldn't think of taking at Y.M.C.A. College in Worcester. Things were very rough in those days. But I find from my correspondence that I was being considered at Northwestern and I was offered a job at Illinois.

. . . I don't know how I could have neglected all that. It makes a good story. . . . It wasn't all that bad."

Skinner devotes full time to writing — no teaching duties or research. He stopped his laboratory work in the mid-1960s, and says, "I always loved it but it always conflicted." Now that he feels he will no longer be making important scientific discoveries, he is concentrating on "what I think I can do best at my age. I think I am in a position to pull some strings together and deal with criticism and the obstacles to progress in the field."

Even without future books, Skinner feels well satisfied. "I plan my life and I think that explains what success I have had. I do not believe that I have an especially high IQ. I think that I have been very stubborn about some ideas and have managed my time well." Incredibly, he finds he is able to say: "If when I was just coming to Harvard in 1928, if I could have prescribed exactly what I wanted to do, and if I had known the present situation, I think I would have been satisfied with exactly what I am now. That is rather remarkable."

What's a Nice Scientist Doing in a Place Like the Press?

Dr. Linus Pauling Reveals in *Midnight* Interview:
  Vitamin C Can Add 8 Years to Your Life . . . And Reduce
Your Chance of Dying by 50%

— Cover Headline, *Midnight* newspaper
September 3, 1973

"When a reporter approaches," Jonas Salk once told a journalist, "I generally find myself wishing for a martini." Like most scientists encountering the press, Salk felt harassed, inconvenienced, misquoted, and misunderstood. A feeling pervades the scientific community that the press is no place for a self-respecting scientist. That today's visible scientists do not share this antipathy for the press sets them apart from most other scientists of the past and present.

Salk appears to have been tortured by the problems of handling the press. A *Look* reporter, who went to Salk's laboratory for a picture story following the announcement that Salk had developed the first successful polio vaccine, recalls, "When Jonas heard what we wanted, he withdrew to his sanctum and didn't reappear for hours. I found out that the proposal of a picture-story was such a trauma to him, coming as it did on top of so many similar traumata, that he coped with it by lying down on his settee and brooding about it. He didn't want to say yes, and he wasn't sure he should say no. He was in a dreadful stew. He'd come out and ask a couple of questions and then go back inside and lie down some more." (Particularly after the establishment of the Salk Institute for Biological Studies in La Jolla, California, Salk of course developed a more sophisticated approach to the press. The National Foundation, which sponsors the annual March of Dimes, has insured Salk's life for $2.4 million, the amount of the asset his name is believed to contribute to fund raising.)

Albert Einstein detested the press, according to biographer Ronald Clark. A month after his theory of relativity made world headlines in 1919, Einstein wrote that the publicity was "so bad that I can hardly breathe, let alone get down to sensible work." Determining to make the best of reporters' "distasteful interest," he answered reporters' requests for pictures by selling photographs of himself and giving the money to the starving children of Vienna.

Scientists exchange horror stories about the press the way laymen discuss their operation scars. Career pressures perpetuate the antagonism: interviews, telephone calls, and press conferences take time from research. Also, a hasty or misquoted remark can jeopardize good relationships with colleagues, employers, and funding agencies. And appearances in the "penny press" can be interpreted as publicity-seeking. A clinical researcher at the University of California Medical School recently asked science writer David Perlman not to mention his name: the last time there had been a newspaper article about him, a couple of colleagues had remarked, "Say, Doc, I saw your ad in the paper yesterday."

Scientists of course have some legitimate complaints about science news, complaints which the corps of trained science writers often share. In fact, the criticisms that come from professional science writers are often more inciteful and sophisticated than those from scientists. Because of the tensions between science and press, and because science is becoming an increasingly important topic for society, science news has become one of the most heavily criticized areas of journalism.

The key concept in most criticism of science news is distortion. The picture of science in the popular press can be like an irreverent political cartoon. Just as a cartoonist plays up a politician's large nose or a president's high forehead, the press sometimes overemphasizes certain features of science: its breakthroughs, its mysteries, its heroes. And like a cartoon, the resulting caricature of science distorts the public's image, changing its impressions of science and scientific priorities. Yet the public, which depends on the press for its science information, uses this distorted image for making important policy decisions. Epitomizing the distortion theme, Philip Abelson, physicist and editor of *Science* magazine, has called most science news "gee-whiz, Buck Rogers distortions of the facts." Donald C. Drake, a science reporter for the *Philadelphia Inquirer*, agreed: "The readers of the American press have a distorted view of the world we present to them."

The most prevalent and hackneyed criticism of science news involves a classic form of distortion: sensationalism. In the process of making a science story newsworthy, the press sometimes transforms a new research finding into a "major breakthrough," a "key to life," or a "cure" for a dread disease. As medical writer Marguerite Clark laments, "While fitting into our copy the classic requirements, the five 'W's — Who, What, When, Where, and Why, how often are we urged by the King's Row boys to add still another 'W' — WOW!" A Detroit editor told journalist Stan Wiggins, "I'll take gee-whiz over scientific significance four out of five times." One scientist who was the victim of a story about a psoriasis "cure breakthrough," and found himself deluged by hopeful patients from all over the world, has vowed never to speak to a reporter or prepare a news release again.

Professional science writers have been among the first to condemn unwarranted sensationalism. Writers point out that some sensationalizing may be necessary to capture the audience's attention: if a story isn't read, it does not matter what it says. But George Dusheck, science reporter for San Francisco educational television station KQED, has said that the greatest failure of the media, especially television, is their failure to report the *tentative* nature of scientific findings. And one journalist nearly lost his job for speaking out against sensationalized violence. In an article that appeared in *Philadelphia Journalism Review*, *Philadelphia Inquirer* science writer Donald C. Drake criticized his own violence-oriented coverage of a speech by Hubert Humphrey at the 1972 convention of the American Association for the Advancement of Science. Drake felt Humphrey's answers to questions from the audience were highly significant, but he, like most reporters, played up instead the demonstration that accompanied the speech — several hecklers pelted Humphrey with paper airplanes. Drake reflected, "I wrote the story that way because I knew, without asking, that that's what the desk would want and I wanted good play for the story. I didn't want to see a wire story substituted, nor did I want to argue with an editor after an exhausting week. And finally I felt that the facts, using conventional journalistic criteria, supported the lead regardless of whether it was the right lead. In effect, I gave them what they wanted. In doing so I became a hack, a whore, not only to my profession but to my paper."

Commenting that a "whore" shouldn't have the latitude of a science beat, the *Inquirer* executive editor demoted Drake to general assignment status under the city editor; after considerable publicity about the punishment, Drake was reinstated in his science writing job.

There is in science news reporting a general need for perspective to combat not only overdone violence but also other exaggerations. There is, for example, a need for historical perspective. Science reporters deal in instant research, a discovery in geology one day, an announcement in biology the next. Because of the press's day-to-day structure and scientists' own efforts to put their best foot forward, events are usually not placed in the context of the continuing process of research in which they occur. The public is led to believe that science proceeds by a series of great leaps forward, rather than by a gradual exploration, with many false starts and failures. New York University professor Hillier Krieghbaum points out that the habit in science reporting of covering finished products is in direct contrast to sports coverage. The sports reporter devotes a great deal of discussion to the pregame suspense, the line-up, the health of the players, past years' scores. In science, nothing is announced until the game is over. Also, it seems that the "discovery" approach of the press is pressuring some scholars to design neat, quick experiments to accommodate the press.

Particularly difficult for scientists is the tendency of the press to spotlight a single individual as the scientist responsible for the discovery or invention. Wolfgang Panofsky, director of the Stanford Linear Accelerator Center, points out that this can be disastrous for the particular scientist in the limelight, who is blamed by his colleagues for not giving them proper credit. Dr. Victor F. Weisskopf of the Massachusetts Institute of Technology recently gave a story to *New York Times* science editor Walter Sullivan about new subatomic particles called quarks. The editor (through no fault of Sullivan) cut thirty percent of the article, including the names of contributors, Dr. Panofsky says, and Weisskopf "got hell from all his colleagues at M.I.T." Jealousies also arise because some areas of research receive more attention than others from the press. Reporters point out that areas of science that are relevant to public concerns must have priority in the press.

Second to sensationalism among scientists' complaints is inaccuracy. Everyone who has interacted with the press for long has an anecdote about being misrepresented or misquoted. Dr. Carl Djerassi, professor of chemistry at Stanford University and director of research for Syntex Corporation, is typical — he says he has "never been quoted correctly."

A comparison of research studies suggests that certain errors, such as overemphasis on the uniqueness of a discovery, are more common in science news than in general news, but that overall accuracy is similar. One problem has been a tendency of scientists to use the same standards for judging the accuracy of newspaper articles as they have for scientific journal articles. Scientists even complain occasionally that a news story did not include enough information to enable them to replicate the experiment described, although this is clearly not the purpose of a newspaper account. Ben Bagdikian tells the story of a social scientist who, after dealing with journalists in Washington for a while, concluded that the trouble was, "you guys don't have any hypotheses."

Third on the list of scientists' complaints is oversimplification. Some simplification of science information is a necessity, to communicate it to a general audience. But the question of how far simplification should go is hotly debated by media critics.

To a great extent, simplification and the resulting distortion are inevitable. According to M. W. Thistle, a scientist who became chief of public relations for the National Research Council of Canada, "To ask a man to translate from one or several rich, relatively new, and precise scientific languages into a single poverty-stricken language of inadequate structure . . . is asking a very great deal. Whatever detail this man does manage to get across to a general audience will certainly be distorted and, to some extent, actually false. No other outcome is possible."

Whether science writers go beyond the inevitable translating process, and in fact oversimplify, depends on who their audience is, or should be. University of Minnesota researcher Percy Tannenbaum has developed a thesis that science writers are good judges of their readers, but that their editors, who have the final say, are not. Tannenbaum concludes from a number of studies that editors are still trying to gear science articles for the "mass," most of whom are nonreaders of science news, rather than for the small, sophisticated fraction of this general audience who are the actual readers. The science readers would prefer less sensationalism and less simplification, which are devices aimed at an audience that ignores science news. It is in the nature of the journalistic enterprise, however, to try to attract the nonreader. A newspaper or broadcasting station is usually a business organization, seeking to increase its audience and thus its profits. Publishers have learned the hard way that newspapers that appeal to elite audiences are usually unprofitable. And in cases where the scientific information helps to clarify a public policy issue, it is really the media's responsibility to attract as many citizens as possible.

Scientists, steeped in a tradition of Ph.D.'s and professionalization, point accusingly to science journalists' lack of science training. "If a sports writer knew as little about baseball," Wolfgang Panofsky commented wryly, "as a science writer knows about science, he'd be drummed out of the corps." Science writers continue to be predominantly journalism or English majors, and editors give higher priority to writing experience than to science training.

The result is a bitter controversy over what the science writer's background should be. Scientists and journalists agree that the science writer needs training in both science and writing, but the question is how much and what kind of each. On the one hand, there is the on-the-job training approach. "Go out and cover a fire," urges Frank Carey. "Learn first to be a good reporter — and never forget you are one." George Dusheck, when asked during a lecture at Stanford University about his science background, replied unapologetically, "I flunked physics." He added that Gladwin Hill mastered the environmental beat for the *New York Times* in two years, with no previous science background. Just as science news has not been a popular outlet for scientists, science reporting has not been a popular job for journalists, and a writer has become a science writer often by default. A young award-winning science writer, Sandy Blakeslee, took a science-reporting job at the *Times* because she was assured it was at the moment the fastest way to the city desk. Although her father, Alton Blakeslee, and her late grandfather, Howard, were widely known science writers, she was surprised to find that she liked the science job. William L. Laurence, the *Times*'s revered former science editor, took the job in 1930 because it was the only one offered; yet he "managed," according to *Time* magazine, "to be something of a scientist himself," often giving scientists leads to new discoveries.

The science writer gets on-the-job training from a number of sources, including stacks of journals, visits to laboratories, governmental background briefings, open houses at industrial labs and similar "junkets," science seminars, science writer institutes, and his "stable of experts."

On the other hand, as the quantity and complexity of science and technology increase, more and more is heard from advocates of formal science training. Professional science writers, although having majored in English themselves, tend to recommend that future science writers take college courses in both science and writing. And younger science journalists have been getting more science training.

The National Association of Science Writers (NASW) lends a certain degree of professionalism, and a sense of identity, to science writing. Newsletters and clipsheets are circulated among over 900 members. And the Council for the Advancement of Science Writing, organized by the NASW in 1959, holds briefings periodically on technical topics of current news interest. Of course, many reporters writing about science are not members of NASW and lack the sense of professionalism it fosters.

Some scientists and most science writers would, of course, like to see more space and time in the media devoted to science news, but competition for the media's scarce resources is intense. The problem is aggravated by the large volume of science information that must be fitted into the media's small allotment of science news. One estimate is that, all told, only 0.01 percent of science information is communicated to the public. Moreover, stories that do get approved for publication are very often cut, reporter William Greenberg asserts, because editors do not realize that the complexity of science demands longer stories. "I can't communicate a science story in less than a thousand words," says *San Francisco Chronicle* science editor David Perlman; "I can communicate a murder story in fifty."

A final criticism, leveled at science news by reporters, not scientists, is the lack of interpretative and investigative reporting. Science writers feel that, given the technological complexity of most political issues today, they should investigate and interpret these issues in their news stories; the "conveyor-belt" approach is no longer enough. They see science reporting as having passed through three stages: first, the gee-whiz, yellow journalism period up until about the 1920s; second, the straight reportorial stage, the conveyor-belt stage; and now, an interpretative stage, in which issues as well as incidents must be handled and analyzed much as James Reston or Walter Lippman would handle foreign affairs news. Implicit in the "three stages" idea is the fact that science journalism has already improved significantly. Frank Carey, reflecting in 1966 on his twenty-five years in science writing, noted that an operation on Shirley Temple's tonsils years ago would have yielded a news story on how she was feeling; today it would be a lesson in anatomy and surgery. But many science writers are still not satisfied. Science writer James Skardon criticized journalists for the lack of investigative reporting in connection with the fire that broke out during a ground test of the Apollo I capsule, killing three astronauts in January 1967. There had been earlier danger signs, such as small fires and safety problems, which the press, as watchdog, should have used to warn the public. Instead, said Skardon, reporters went along with the "NASA myth of invincibility," providing "uncritical and even idolatrous coverage."

Unlike political reporters in the Washington press corps, science writers have traditionally not been adversaries of the institution they cover, not inclined to dig out a Watergate within science. Because science has seemed to be a dull topic which needed to be glamorized and "sold" to the public, science writers have functioned more as good public relations men for science than as traditional, watchdog journalists. In the past ten years, however, there has been an increase in coverage of unfavorable and interpretive aspects of science, probably partly because science writing has become more professional, and partly because science has become more pervasive, more powerful, and more political. "These days," according to Victor Cohn, "it is not enough for us to report the new discoveries and gadgetries; we must delve deeper into their effects on people and public policy . . . be a public watchdog."

The result of scientists' problems with the press has been what Krieghbaum calls a "Yes, but . . ." reaction: *yes*, the public should be informed, scientists say, for our good and theirs, *but* the way science reporting is handled is deplorable. Within the "yes" is scientists' acknowledgment that the public has a right to know, and that providing information to the press may increase public favor and hence research funding. Also, many relatively unknown scientists are flattered, of course, to have a little attention from reporters. Within the "but," however, are all the career pressures, personal inclinations, and political realities weighing against speaking out in the press.

Scientists' antagonism toward the press has been further aggravated by their own ignorance about the press and about the process of communicating with the public. As a result, scientists have been partly responsible for creating some of the very problems in the media they deplore.

One of the most obvious ways of creating havoc has been by

failing to cooperate with the press. Whether out of idealistic disdain, previous traumas, or fear of reprisals, scientists often cut off reporters with a frosty "no comment." While it is understandable for the scientist to lick his wounds in silence or concentrate his efforts on the laboratory, his unavailability causes serious problems. A vicious circle is created, for as the scientist withholds information, the reporter's story becomes increasingly inaccurate, the scientist even more bitter, on and on. Also, the inaccessible scientist ensures that the information he possesses will be poorly presented, and that the more talkative scientist will receive proportionately more coverage than he deserves.

The silence of scientists has particularly severe repercussions in policy areas. In his documentation of the events surrounding the appearance of Rachel Carson's *Silent Spring*, Frank Graham, Jr., blames the academic scientists ("silent scientists") for the breakdown in communications that delayed public awareness of the pesticide crisis: "In the face of man's massive intervention in the functioning of the natural world, the scientific establishment simply filed the ominous facts and kept mum. . . . They sneered at such techniques as 'popularization,' and recoiled in indignation from the suggestion that they cooperate with the mass media to put across the story that should have been told." Scientists in government advisory positions have been the most likely to possess important policy information and the least likely to speak out. Project Argus was an experiment in which nuclear weapons were detonated above the atmosphere. With the cooperation of the members of the President's Science Advisory Committee and other scientists, the project was kept secret, for fear of alarming the public at home and creating diplomatic tensions abroad. A few reporters gradually investigated, and in March 1960 a story announcing the secret tests was published in the *New York Times*.

The scientist further stymies the press in many larger research organizations with an elaborate security system. The reporter is required to go through an information officer, who "protects" the scientist and the organization from rash statements the scientist might make. At the Stanford Linear Accelerator Center, Douglas Dupen, head of Public Information, tries to handle all reporter questions himself. Rather than refer technical questions to the researcher involved, Dupen acts as mediator, or at least he tries to be present if a meeting between scientist and reporter occurs. Science reporters covering the space shots found that NASA had particularly stringent security precautions, including elaborate procedures to ensure that a representative of the Public Affairs Office was always at hand during an interview between reporters and NASA employees. Such "security" measures protected NASA from revealing not only classified information but also highly embarrassing facts, such as the existence of dangerous flaws in the space flight equipment. Other science writers have encountered cases in universities in which a professor was not allowed to speak to a reporter without administrative clearance.

Another form of scientific censorship comes from the norm among scientists that research developments must be reported in technical journals before they reach the popular press. This norm is so strong that some journals refuse to publish articles whose contents have been previously announced in the public media. *Physical Review Letters* set a precedent in 1960 by announcing that it would reject papers whose main contents had already appeared in the daily press. In 1970, as editor of the *New England Journal of Medicine*, Franz J. Ingelfinger put forward a similar philosophy, and the policy became known as the "Ingelfinger rule."

This publish-in-a-journal-or-perish rule protects the system of quality control in science. Before publication, a journal's referees judge the acceptability of a manuscript for publication. After publication, an invisible college of colleagues reviews and evaluates the research, and censors inaccurate findings before they can mislead future researchers or the public. If the discovery appears first in the popular press, the discoverer's colleagues cannot perform this function, since the popular article does not have enough detail for them to judge its merits. Reporters cannot check up on a story as easily, because their sources have no previously published technical articles on the subject that they can use for reference. The journal system also discourages charlatans who might make unfounded announcements in the press for publicity; stories in the popular press are not awarded scientific recognition and in fact, often receive disapproval.

But this practice also ensures that the public will get a "voilà" view of science. They will see the approved final product, not the trial-and-error experimenting, debating, and groping that make up the scientific process. In their concern to protect people from substantive inaccuracies, scientists mislead them much more seriously about the essence of the scientific process. Also, by stripping science of its natural human interest, they leave it for reporters to add artificial drama. In short, scientists contribute to the very sensationalism and "discovery" orientation of the press that they criticize.

Moreover, scientific censorship does not succeed in protecting the public. False claims of cancer cures still appear. "Caveat emptor" has been the rule in most consumerism; perhaps it is time the public realized it must ultimately judge, also, the goods from science and technology. The reader, John Lear argues, can be trusted to exercise some judgment of his own about tentative hypotheses. It is obvious from research on audience barriers that readers do not soak up news from the media indiscriminately. A few inaccuracies in the substance of scientific articles may be well worth tolerating to avoid a general inaccuracy in the public's impression of the workings of science. As Victor Cohn put it, "Neither journalist nor scientist owns science; it is in the public domain."

The publish-or-perish system also generates a prodigious volume of technical literature, which in itself acts as a barrier to public communication. The press is deluged by too much, too obscure material. And the volume of technical literature has been doubling every ten years. Science reporters simply do not have the time or technical competence to sort out important developments from the piles of print. Increasing the volume of distracting information, according to Cohn, are the public relations offices of universities, government agencies, and companies, whose "guilty, mad mimeographers" generate exaggerated news releases in their eagerness to generate good publicity for their organizations. In many institutions, scientists are given an opportunity to screen these releases, and sometimes they encourage the huckstering, helping to sell themselves, their institution, or the scientific enterprise.

While some of the science community's public relations efforts are professional and effective, many scientists fail, even when they wish to communicate with the press. Sometimes this is caused by a failure to understand what makes an appropriate news story; sometimes it is plain ineptitude. Scientists are notoriously poor writers, not only in materials intended for the lay public, but even in their own technical journals. Their failure to learn the techniques of communication results, Thistle says, in "some of the worst-written documents in the world."

Ironically, then, the press is plagued at times by too little information from scientists, and at times by too much. In part, the paradox is undoubtedly a reflection of the conflicting values in science about public communication. Interestingly, scientists have not always had a reputation for being obscure communicators. In the seventeenth century, scientific writing was apparently considered simple and pure, reminiscent of the common man's speech, in contrast to the complexities of literary style. The purpose of the Royal Society in London, according to Thomas Sprat, a seventeenth century historian, was to "return back to the primitive purity, and shortness . . . a close, naked natural way of speaking . . . as near the Mathematical plainness as they can: and preferring the language of Artizans, Countrymen, and Merchants, before that of Wits, or Scholars."

Scientists are of course not the only problem in science news; they are the first in a long line of barriers to communication, as science information circulates among scientists, public relations officers, reporters, wire service editors, news editors, and publics. Since the ethos and aims of science have conflicted with the values and objectives of the press, tensions and hostilities have built up between scientists and reporters, which in themselves inhibit the dissemination of news. The tension tends to decrease communication between the two groups, which decreases the quality of science news, which in turn increases the tension.

Jealous of their right to pursue science in peace and freedom, scientists have resented the intrusion of the press and its threats to the system of technical publication. Upholding the public's right to know and the press's role as a watchdog, reporters have resented the lack of cooperation from scientists. Dramatizing the reporters' point of view, a Canadian university public relations director remarked, "After fifteen years in this job I'm thoroughly persuaded that if there is an information gap it's the fault of the whole g-- -----d [sic] science establishment, its privileged status, lack of public accountability and ivory-towered complex: it will talk when it is damned well ready to, to whom and no sooner. And the rest of society, who pay the shot, should be grateful to have them on the payroll."

In some ways, professional science reporters are middlemen, charting a course between scientists and laymen. They appreciate the means and ends of science, and also appreciate the layman's difficulty in understanding science. They acquire a general background so that they can decipher the jargon and culture of the scientist, but they also acquire journalistic skill, an awareness of the layman's requirements for comprehension. In covering science, they get to know scientists and to sympathize with some of the scientists' complaints about the press. And they share some of the scientists' frustrations over misleading headlines, stories edited too heavily, and other aspects of the news process over which they have no control. Sometimes they find themselves writing more to please the scientist than the layman. But they view themselves as journalists and not scientists. They have an allegiance to the goals of the press, and an understanding that science stories must conform to the standards of journalism, not of science. As a result of the schizophrenic quality of their job, science reporters clash with scientists, when they carry out their function as newsmen, and with their editors, when they seek to improve the quality of science news. A prominent newspaper writer considered himself an "intermediate censor," for example, when Ehrlich and Commoner were at loggerheads in 1972. Both environmentalists provided the writer with sensational material on their quarrel, which both they and the newspaper would have liked to see printed. The writer never used the material, however, believing that Ehrlich and Commoner were too valuable for the public to find out what "boobs" they were.

While it should not be a scapegoat, the press itself must of course take some of the responsibility for the problems in science news. Most theorists and critics of the press today would agree that, in return for its freedom, the press has a responsibility to the society it serves. With respect to science, editors, not scientists, generally decide what stories are covered. It is the responsibility of the press, then, to respond to the public's interest in science, and to present in-depth treatment of the scientific aspects of policy issues.

Although the reporter writing a science story may have considerable sophistication about science, many important decisions are out of his hands. City and news editors, who are unlikely to have science training, determine what stories will be assigned and used. Deskmen and copy editors edit, cut, and headline the stories. Public relations offices control the issuing of news releases. And regional and local wire service editors determine what stories pass along the wire and into the regional newsroom. Sensitive to the problem of wire service editors, Alton Blakeslee, the Associated Press science editor, has been quoted as saying that he could get a story on the front page of every newspaper subscribing to Associated Press. All he had to do was to mention in his lead paragraphs something about a treatment for piles, ulcers, or sexual impotence. Every telegraph editor has these three conditions or worries about them, Blakeslee explained.

The process by which stories reach reporters is also erratic and unreliable. Stories usually come to a reporter's attention through a scientific meeting or convention, where the reporter depends on "corridor consultation" to learn which announcements are important. Otherwise, public relations bureaus in universities and industry send out releases promoting their particular institution. And occasionally the scientist himself will raise an item with the press, leaving it to the reporter to find colleagues to verify or refute the scientist's claims. In all these cases, the reporter must depend for advice on someone who has an ax to grind. The selection process may also be further hampered by "pack reporting," a tendency for reporters to copy each other, covering the same story already singled out as important by a colleague on another paper.

In addition to scientists, reporters, and editors, a fourth group creates problems for the communication of sciences: the public itself. Communication theorists no longer believe that the mass media, like a hypodermic needle, inject information directly and efficiently into men's minds. What the media make available and what the public learns are two very different things. To begin with, by no means does everyone read science news. Ironically, it is those who already know something about science, and have favorable attitudes toward science, who are most likely to read science news. The typical science news consumer is fairly young, an urban dweller, well educated, in a higher income bracket, a heavy user of the media in general, and a former student of science in high school and college. In other words, many people are cut off from science news simply because they do not read it. In addition, once a science story finds a willing audience, it falls victim to each reader's psychological mechanisms for selecting, rearranging, and distorting the information he receives and remembers.

In general, people read and remember science news because it is news, not because it is science. Studies show that people seek more and learn more from the media on science topics that they find relevant to their lives and interests, and that the public is interested in science for its inventions, implications, and news value rather than for the importance it places on basic research. According to a University of Missouri study, most news editors, unlike science reporters and scientists, do not distinguish between science and other news, noting rather that many aspects of day-to-day problems, from weather to farming, are science related. In order to identify and address special problems, some writers and researchers have taken science news captive and isolated and examined it; now it is time to toss it back and see how it fits with the rest of today's news.

Enter the Visible Scientist

Amid the traditional troubles and tensions in science journalism, the visible scientist is an anomaly. He welcomes the press, understands its workings, tolerates its failings, and feels at ease in its activities. In fact, some of his attitudes may be more congruous with reporters' than with most scientists'. Like the media, the visible scientist is working toward a better flow of science information to the public, as distinguished from the flow among scientists or among laymen.

The relationship between science reporters and visible scientists is symbiotic — each is dependent on the other. Science reporters are so dependent on scientists for their stories that they are often described as parasites in the news process. But in the case of visible scientists, the reporters are using scientists who want to be used. Visible scientists want publicity for their issues and ideas. Direct, quotable, and newsworthy, their messages stand a good chance of escaping some of the typical censorship and editing of science news.

One finds instances among visible scientists of remarkable cooperation with the press. B. F. Skinner, during a long interview at his home with a free-lance writer, found it was nearing time for him to go out for the evening. He invited the reporter to remain in his basement office and peruse his personal diaries. Skinner assumed — falsely, he now feels — that the writer would ask his permission before printing any of the personal material. In fact, an introspective passage was quoted in *Harper's*. Margaret Mead has been known, friends say, when a floundering young reporter is assigned to write about her, to write the story for him. Even Glenn Seaborg, who had frequent contact with the media when he was chairman of the Atomic

Energy Commission, said he got to know some reporters in Washington "almost as though they were friends."

Most of the visible scientists are generous with their time, giving long interviews and copious material about themselves and their work to newsmen — too much at times. Environment reporter Eliot Porter laments the existence of "excess-response men," scientists who deluge reporters with too much material, and pile on extra reading matter, as the reporter backs out the door.

Visible scientists are remarkably tolerant of the press's failings — not blind to them, but not bothered. They give the press no prizes for accuracy, but they consider the overall quality of science reporting good. They have come to expect, and accept, inaccuracies in news stories. And they handle the problems with sophistication and perspective.

René Dubos recalls giving a speech in New York in which he suggested that the two-child family is not the solution to the population explosion. Children are traditionally raised in large families, he said, and the small family could lead to emotional difficulties; for all he knew, it would be better for some adults to remove themselves from the reproductive pool, while others had several children, averaging out to zero population growth. The headline on a resulting news story read: "Rockefeller Scientist Says Two Children Not Enough." "I don't have to tell you," Dubos says, "what that brought down on me."

Dubos's reaction to the incident was sophisticated. He noted that it was only the headline that was misleading; the text of the story was fine. And the headline, he explains, was written by a different person from the reporter who covered the speech and wrote the story. Paul Ehrlich adds that he puts up with the problems because "I have no choice." He seldom fusses about retractions, adding a comforting note: "There is nothing so dead as yesterday's news; no matter how hideously garbled it is, it's all over."

When visible scientists assign blame for science news problems, reporters usually get the benefit of the doubt. Headlines are a common complaint, and always with the recognition that they are not written by the reporters. Said Joshua Lederberg, "Sixty percent of the problems are the headlines, and what the scientist at large doesn't know, what the public doesn't know, is that the guy who writes the headlines is not the guy who wrote the article, and he [the reporter] is just as furious as anybody else." Visible scientists also recognize that the reporters' stories are changed by editors above them. "I have a feeling the major filters are higher up," Ehrlich remarked; "the bigger mistakes are made in the decisions of what is to be carried, further on down the pike."

The visible scientists' praises, however, are reserved for the professionals who "know their stuff." Just as certain scientists acquire a reputation for being good sources, certain reporters acquire a reputation for being good science writers. Word gets around about the outstandingly good ones, and the outstandingly poor. The reputation of a science writer among scientists is crucial to his job. A good reporter develops a "stable of fans," as one writer put it, scientists who trust him and upon whom he can call. David Perlman, science editor for the *San Francisco Chronicle*, was praised, for instance, by nearly every scientist interviewed on the West Coast.

More surprising, the visible scientists often blame themselves for stories that misrepresented and misquoted them. Understanding the reporter's problems, they feel it is up to themselves to be more careful — if a remark has been misinterpreted, they should have made their point more clearly. Margaret Mead explains, "I expect small inaccuracies. If there are major inaccuracies, I usually consider that I made a mistake." Citing an example, Mead described the background behind a recent flurry of stories in the press that she favored legalization of marijuana. During Senate testimony, she was asked if there was too much use of pharmaceuticals in this country: "I simply gave them a lecture on the whole attitude towards drugs in this country, and in a subclause I said that legalization of marijuana would help some but not much. That's all I said. But I did use the phrase 'legalization.' . . . I know enough never to say 'legalize' anything people disapprove of. You see, you don't say 'legalize' — in the United States that means sanctify. I've never made that mistake with abortion. I've always said, 'No, I don't believe in legalization of abortion; I believe in repealing the laws against abortion and returning it to church and the medical profession.' I made a technical error. Now, I think I'd have been bitter at the press at that point, if I hadn't made an error myself."

Even when the error is clearly the press's fault, Margaret Mead says, "I don't get mad at them. Most people get mad for ridiculous reasons." Instead, she tries to be careful, and discriminating, in her contacts with the press. "Most people don't pay enough attention. They are busy doing other things." She recalls a time when she refused to finish a radio interview for one of the major networks because she found out the session was being taped while she was questioned by one interviewer, then was going to be edited and fitted to questions interposed later by a national commentator, Bob Considine. Afterwards, one of the men who had been operating the recording equipment shook her hand and congratulated her for her decision. Others, she found out, did go ahead with interviews and were shocked to find that what they said was "chopped up in little pieces."

Mead also knows, in the reverse situation, why, when she wants to get something in the press, it may not appear, At the time of the furor about her remarks on marijuana, she taped a television response that was never used because of a plethora of other stories breaking that weekend. "I know how journalism works. I know if you die at nine o'clock in the evening you will never have an obituary in the *New York Times*, no matter who you are, virtually. You might rate the front page, but not an ordinary obituary."

Not all the visible scientists have the sophistication of Margaret Mead in their dealings with the press, but, like her, they learn from their mistakes. B. F. Skinner recalls that the first time he was ever on a television talk show, "something came up and I raised the question — which had bothered me — a question raised in the first place by Montaigne. He asked, 'Would you, if you had to choose, burn your children or your books?' Montaigne said he would burn his children. I had talked with both my children, saying, 'I would burn you, too; I'd burn myself. I believe that my contribution to the future through my genes is not as great as my contribution through my work. And I just want you to know that's the way I feel. Fortunately, I don't have to do this.' Well, I brought this up on the show, and you can imagine the reaction!" Although Skinner still feels his answer was correct, and he has explained the issue to his two grown daughters, he has learned to avoid such inflammatory statements in the media.

Talk shows are trying experiences, even for visible scientists. Typically, visible scientists have appeared on talk shows — "Today," Johnny Carson's "Tonight" show, William Buckley's "Firing Line," and others. And the scientists have not been impressed. The talk show host does not do his homework, they say. Instead, he gets off the subject, and centers attention on his own antics. Barry Commoner recalls one who confused him with Paul Ehrlich and asked about his vasectomy. You learn, Jane Goodall says, to brush aside a stupid question, and with a "that reminds me" proceed to say what you wanted to say.

But visible scientists learn to discriminate among talk show hosts, as they do among reporters. B. F. Skinner had a close call a few years ago when Zero Mostel asked him to appear on the Dick Cavett Show. Skinner had known Mostel for many years and they both vacationed on the same island in Maine. "He is a very intelligent man, and I thought maybe he wanted to make a name for himself as that type of talk show master of ceremonies," so Skinner agreed. The night before he was scheduled to appear, however, Skinner and his wife watched the show. "Zero poked his head in and mouthed the audience and hammed it in the crudest way. The first fellow he brought on, I felt sorry for him. . . . Zero was just murdering these people for the sake of gags. . . . He started talking about *Portnoy's Complaint*, trying to get these people to describe their sexual [hang-ups]. . . . Well, we had been watching this for about ten minutes and I turned to Eve and said 'I can't go on.' She said, 'No, of course you can't.' So I said 'What will we do? Do we tell Zero, or do we give him an alibi for NBC? . . . We'd better give him an alibi.' So the next morning she called NBC that I was down with the flu."

The highly visible scientists tolerate the inconvenience and close calls largely out of a sense that television is a very powerful and important medium. According to Margaret Mead, "Television is the best medium. It is not a good medium for a long substantive communication, but the more you are on television, the more others pay attention, read your books, come to your lectures. There is nothing like it, because television appears to be extremely frustrating. Americans want to be in the same room with people and they don't feel that television puts them in the same room. So if they have seen you on television, they read your book or they go to a lecture. They drive fifty miles to sit in the back of a hall with two thousand people, because they have seen somebody on television. . . .

"There are fifty employees in this building," Mead added, concerning her apartment building on Central Park West, "and they change all the time. But when we moved in here, our first job was to get to know fifty employees — so I went on the Johnny Carson Show right away and then everybody knew me."

Margaret Mead and the Media

Margaret Mead, remarkably sophisticated about the press, is probably the most visible contemporary scientist in America. Filling one wall of an elaborately decorated restaurant in Harvard Square, now called "33 Dunster Street," there is a row of giant stained-glass windows, each a modern icon, in mock-saint costume. The enshrined figures, erected in early 1972, are Richard Nixon, Spiro Agnew, Joe Namath, Bobby Orr, Humphrey Bogart, Germaine Greer, W. C. Fields — and Margaret Mead. "If a non-anthropological audience is asked to name a prominent anthropologist," writes an anthropologist who studied Mead for his master's thesis, "the first name suggested will almost without fail be that of Dr. Margaret Mead. In my experience this will be the case on every level from high school students to academicians in other fields." Adds British anthropologist Peter Worsley, "There is little doubt that the American and British reading public gets its knowledge of anthropology almost exclusively from the works of Margaret Mead."

Margaret Mead has been visible in varying degrees since 1928, when her first book, *Coming of Age in Samoa*, became a best-seller, taking everyone, including Mead, by surprise. "I had no idea it would be popular," she recalls, "and I didn't know it was a success for six months after it was published, because I was in the field again," in New Guinea.

For the public and the press, Mead had a remarkable and successful combination of ingredients: she was readable, relevant, colorful, controversial. Journalists were quick to appreciate these characteristics, which have matured and magnified over the years, as she continues to lecture, write, and speak out in the press on a variety of issues, changing topics with the changing times. A science writer in 1973 summed up Mead's powerful appeal to journalists: "If I went to a scientific convention, and I spotted Margaret Mead, I'd go up and start interviewing her. Margaret Mead is positive feedback, a household name; she's good copy; she's a sure thing."

One of Mead's most important characteristics for journalists and laymen is that she writes and speaks "in English," as she puts it — no jargon. A *Time* magazine writer described a recent lecture as "packed with provocative opinion, and necessary forays into social science jargon were leavened with literate wit."

Margaret Mead did not originally set out to reach either the general public or the press with her readable style. At the time she wrote *Coming of Age*, she made a conscious decision to write "in English," not for the public, but for a specialized group that she felt would be interested: teachers. She explains, "My first book was about adolescence, and it seemed to me the people who were interested in adolescence were not primarily anthropologists but educators, and educators were lay to anthropology. That is, every group is lay to any other specialized group. So it wasn't a question of writing down; it was simply a question of writing in English. In those days, a good proportion of anthropology was utterly unintelligible. It was written with a mass of native words in it so it looked as though it had measles. Malinowski used to put things in Latin that other people didn't put in Latin, to make them look pornographic — like facetiae, that sort of look like feces. I simply decided that the point of having done the study was to tell something about adolescence, and who was interested in adolescence? — Teachers."

When she returned to New York in September 1929, after a year in New Guinea, friends greeted her with the news that she had achieved a wide reputation as an author-anthropologist, "something I vaguely knew but had not yet fully comprehended." The success of *Coming of Age in Samoa* seemed to determine her future direction, a direction that was to bring her increasing visibility. She wrote her second book, *Growing Up in New Guinea*, also in "English," and was more prepared for it to be popular. "I had set a style by that time and I was writing in English and I had decided writing in English was a good thing to do." She also began writing articles for general magazines, from *Parents* to *Nation;* she appears in a broad spectrum of magazines today, and has a monthly column, coauthored by colleague Rhoda Metraux, in *Redbook.* Her position at The American Museum of Natural History in New York has given her an additional forum for popularizing anthropology. She has been with the museum for fifty years, starting as assistant curator of ethnology in 1926, becoming associate curator in 1942, curator in 1964, and curator emeritus in 1969. Her work at the museum culminated in the opening of her exhibit, Hall of Peoples of the Pacific, in May 1971.

Mead's interest in popular writing reaches back into her childhood. She began writing poetry when she was nine, started a novel in grammar school, wrote short plays for school, and helped start a school magazine. In her autobiography, *Blackberry Winter,* she recalls "Writing was what my parents did [her father was a professor of economics, her mother trained in sociology], and writing was as much a part of my life as gardening and canning were in the life of a farmer's daughter of that day." When she started college at DePauw University in Indiana, she intended to be a writer, and when she transferred to Barnard College in New York City a year later, she continued to major in English. "Billy Brewster, with whom I took Daily Themes," she adds, "said I would never be a writer."

As late as 1932 she published a poem in the *New Republic,* entitled "Absolute Benison." It was the last poem she published, she explains in the introduction to a bibliography of her works, because she first submitted it under a pseudonym and it was rejected, then she resubmitted it under her own name in 1932 and it was accepted. Increasingly, she has found the most satisfying channel for her writing interests in her books, on anthropology, in "English." "I have really not written poetry that mattered very much since I wrote *Sex and Temperament,*" she says. "*Sex and Temperament* seemed to me a satisfactory literary form. In other words, it said what I wanted to say. I have written a few poems since then but only when I wanted to say something to somebody, not as a desire to write poetry." Reviewers have often noted that her books were rather like novels, more literary than reportorial in style.

Mead has also exploited an uncanny ability to give lectures that are extemporaneous and captivating. "I taught myself to speak, under my father's watchful eye as he sat in the audience (I was making extemporaneous speeches while I was in High School). . . . My first paper, at the British Association, was the only paper I ever read — I was twenty-two." Her technique is to speak from memory, without text or even lecture notes, on topics about which she has studied in advance. "I don't speak about things I've written about," she says. "I speak to find out how to write them." She never gives the same lecture twice, although speeches may be based on the same themes, constructed from the same building blocks. Remarkably, she generally has "no anxiety as a public speaker," but if she is paid a high fee, she worries about her clothes, "because the program committee is likely to be criticized."

The effect is a spontaneity to which Mead's audiences quickly respond. She has a gift, a graduate student has said, "of being able to capture the thing that everybody knows but rarely says. She can say it in a few words. She hits at the very human truth which is familiar, but which has never been said. And somehow that's very funny. It's not a joke — but it makes you feel warm."

In lectures and interviews, Mead has an irresistible quotability for journalists. Articles about her are peppered with quotations of off-the-cuff remarks, sometimes featured in specially marked-off sections or on separate pages of "Meaty Meadisms." Her arguments are refreshingly clear of qualifiers and disclaimers, like "possibly," "one solution I might suggest," "the evidence tends to point to," "my opinion on that would be." When she answers a question, it is obvious the reporter is asking for her opinion, and that she is giving it; she simply makes her point:

Q: There is a lot of emphasis today on sex techniques as a way to marital happiness. Do you think this approach has validity?

Mead: Well, it isn't a matter of validity. It either works or it doesn't. Until about ten or fifteen years ago, most Americans were appallingly poor lovers. Then they began reading about it and things changed . . .

Other magazine quotes on American family life:

Motherhood is like being a crack tennis player or ballet dancer — it lasts just so long, then it's over. . . . We've focused on wifehood and reproductivity with no clue about what to do with mother after the children have left home.

Different cultures and different periods have different styles of sex relationships. . . . You have the American notion in which the boy and girl are supposed to be of the same age and equally ignorant and clumsy. . . . No one has the slightest proof that very many girls in the United States, age 16, are *enjoying* sex. Not the slightest proof. Why should they, with the partners they've got?

Women are much fiercer than men. Nobody has ever given us weapons for very long, have they? Unlike men, women fight protectively. Women play for real. To win! . . . The ferocity of the girls in the activist groups is far more horrible than the men's. They kick below the belt.

For the first time in human history, there are no elders anywhere who know what the young people know.

Fathers are spending too much time taking care of babies. No other civilizations ever let responsible and important men spend their time in this way.

Mead is a natural for the "People" sections proliferating in news magazines; one squib recently quoted her beliefs about UFOs: "Of course I believe. Do people believe in the sun or the moon or the chairs they are sitting on? We can only imagine the purpose of these visits, but it seems to me most likely that a responsible society outside our solar system is keeping an eye on us to see that we don't set in motion a chain reaction that might have repercussions far beyond our own sphere."

Mead's willingness to defy critics and grapple with contemporary and controversial issues has also made a strong impression on journalists. Writes David Dempsey in *The New York Times Magazine:* "Unlike most of her colleagues, who bury themselves in the tribal customs of primitive man, Mead is visible, a willing plunger into modern social controversy who projects herself as a global prophetess on almost every subject that concerns the human condition. To a society troubled by its own shifting folkways, and hungry for guidance in coping with them, she is a poor man's anthropologist."

Like her decision to write in "English," her choice of topics for field work was originally made with no thought of their popular appeal. As a senior at Barnard College, she had taken a course from Franz Boas, the "father of modern anthropology," and during the course of the year she settled on anthropology as a career. Boas directed her graduate work at Columbia University, and in 1925, when she finished her doctoral dissertation (an analysis of cultural stability in Polynesia), he had definite ideas as to what field work she should pursue. He wanted her to work with American Indians, so the work would not be too dangerous, and he wanted her to study adolescence. Mead wanted to go to Polynesia, and to study culture change. They compromised on Polynesia, but studying adolescence. They agreed on an island to which a ship came regularly, Tau, American Samoa.

To Mead, examination of contemporary problems is also a legitimate role for the anthropologist, and it is one to which she has consciously devoted a portion of her time. By 1935, she reminds people, "I had what amounted to a lifetime of completed work behind me." When World War II broke out, she wrote her first book on American culture itself, *And Keep Your Powder Dry*, and a substantial body of work on national character and industrial society has followed. Since 1940, Mead has divided her time, as she analyzes it, among teaching duties (graduate and postgraduate students), writing scientific articles and reviews, field work (she took her seventeenth field trip in 1975), participation in scientific and lay conferences and symposia, work in scientific societies, and writing for the general public.

Mead has been able to stay relevant and current by adding to her repertoire a broad range of subjects that she studies and comments on — environmental systems, population control, women's liberation, nutrition, violence, black power, drugs, primitive art, cybernetics, the generation gap, the military, tribal customs, urban planning, sex, alcoholism, the church, architecture, civil liberties, human brain evolution, and child development. Uncannily, she often begins talking about an issue, such as population control or urbanization, just before it breaks as a popular topic in the press. She attributes her foresight to a kind of world view. When she visits a country, for a conference or meeting, she says, "I pay attention. I think I see more of what's happening. Most scientists most of the time are pretty busy with their own material, and they go to another country, meet the chemists there or the physicists there, and they don't look at the whole country."

The very fact that her field is anthropology gives her a head start in relevance to the media, Mead believes. "My material is so much more usable. I'm discussing my own field; I'm discussing people." When CBS used to do a show at The American Museum of Natural History, the television people always ate with the anthropologists, she recalls. "We are more like media people." While much of her work is within her own discipline, she points out, too, that much has been interdisciplinary, which is likely to be even more intelligible to the press.

When Mead has a new idea, she looks for a scheduled lecture where she can discuss it, and the press takes it from there. "I am interested at the moment in ectogenetic memory — how we are going to construct one," she said in a 1973 interview. "I have to give the C. F. Menninger Lecture in Florida, and I thought I would put the ectogenetic memory in that lecture, but then when I looked at it again, the lecture was supposed to be on some aspect of adolescence, so it can't be done that way. I said, 'Okay, the next lecture I am giving is for Barry Commoner's group in St. Louis and it's completely relevant.' So I have moved ectogenetic memory to that. Then somebody will ask to publish it, somewhere. They usually publish things seven places; that is one of the horrors of life at present. They publish and republish it, and then translate it and publish a new edition; and you have to write an introduction to that, and then they want to cut out two paragraphs, and then they want illustrations, and this just goes on and on and on indefinitely."

The only audience Mead makes no effort to reach is "the snob audience." "The last time I encountered *Harper's*, they wanted to take a chapter of *Male and Female* [1949], which they said I would have to rewrite: four editors took me out, to explain how their readers couldn't understand it. The *Ladies' Home Journal* then took it, without any change of vocabulary, and *their* readers could understand it. So I have nothing to do with snob magazines."

Adding to Mead's attractiveness as a journalistic subject has been that magic ingredient: controversiality. The very characteristics that have endeared her to the public and press — the relevance of her remarks and the clarity with which she makes them — have had the opposite effect on some of her fellow anthropologists. And journalists have been quick to cite her caustic critics. *Time*, for example, has quoted an injured colleague as remarking that when Mead attacks the wrong-headedness of a fellow scholar, "she is truly like one of those terrible Indian goddesses, standing on her victims with her tongue sticking out." It has been said of anthropology, one critic reflected in an interview, that "to publish a popular book is like a woman losing her virginity: after that anything she does is suspect." There is among anthropologists a feeling of discomfort when Mead's name is mentioned: discomfort that she is popular, and then discomfort that the popularity should bother them.

The solution for some of her colleagues is not to take her seriously — or rather, to take her popularity seriously, but not her work. "The professional anthropologist," according to Peter Worsley, "is liable to react to the mention of Margaret Mead's name with, at best, a smile, and probably with some more positive expression of distaste. Yet few of them have attempted to analyze her work, or to make it clear exactly what it is they object to." Critical evaluations of her work are found almost exclusively in reviews of her books in technical journals; more general criticisms are confined to off-the-cuff remarks at anthropological conventions.

Taking aim at her books written for lay audiences, critics describe Mead's style as "impressionistic," "glib," "melodramatic." Her approach is said to be "intuitive," not based on sufficient empirical data and objective observation. Particularly bad, say traditional anthropologists, are her intuitive leaps from the primitive cultures she is studying to the modern cultures whose problems she seeks to improve; in drawing analogies and making comparisons between primitives and moderns, she leaves the realm of anthropology. The problem is more severe, critics say, because she seeks not only to comment upon society but also to change it. Technically, Mead's interpretations of the data, as a member of the "culture and personality" school, which puts more emphasis on the individual personality in shaping events, are criticized by anthropologists inclined more to the "structuralist" school, which deals more in institutional and historical generalities.

There is also the ever-present criticism that Mead has concentrated on topics related to sex in order to gain popularity. Says a typical critic, "Though she tackles serious questions, there is little doubt that she has played upon the fact that many of her readers have a less than scientific interest in sex and in her work, aroused by titles and headings suggesting all sorts of salacious possibilities — 'Fathers, Mothers and Budding Impulses' . . . 'Women, Sex and Sin.' " Mead heartily resents critics who say she is popular because she has written about sex. Other anthropologists, she points out, also wrote books about sex, such as Malinowski's *Sex and Repression in Savage Society*, with less popular success.

An extension of the criticism of Mead's use of sex as a topic is the accusation that her work is sensationalistic and distorted in general, tailoring the facts to fit a neat hypothesis or a convenient literary form. At the time her first book, *Coming of Age in Samoa*, achieved popularity, anthropologists speculated that she had gone to Samoa to prove a point, to show that the stormy, stressful, American-style adolescence was not inevitable, that the Samoan adolescents had no such difficult period. Suspicions were even stronger when her third book, *Sex and Temperament*, showed three neatly contrasting types of tribes; in her Preface to the 1950 edition of *Sex and Temperament*, Mead calls it her "most misunderstood book," and denies that the "too good to be true" pattern was anything but the way she happened to find it when she went to New Guinea to study. Peter Worsley summarizes anthropologists' criticisms: "The aggrieved specialist therefore condemns her writings as the work of the 'rustling-of-the-wind-in-the-palm-trees school'; and he condemns the vagueness of the information, the plethora of irrelevant atmospherics, and the stamping of the material into Procrustean molds. At the worst, this becomes downright distortion."

Mead answers her critics by pointing out that they base their attacks on her popular books, not her technical work. "This makes them feel guilty, because they know that I have written an awful lot of technical stuff, some of which they wouldn't understand if they read it. And they don't read it. They read the things that are written so that they are intelligible to bright freshmen." They think they have gotten all she has to say from her popular stuff, according to Gregory Bateson, an anthropologist and her former husband; he agrees that anthropologists do not read her technical articles. In fact, in many cases critics seem unaware that they exist.

Mead is justifiably concerned that the press's emphasis on her critics gives a distorted impression of her standing with her colleagues. While the critics make colorful copy, there are a number of indications that Mead's work is held in high regard by the majority of anthropologists and other scientists. She has held important positions in scientific societies, serving as president of the American Anthropological Association in 1960, the Society for Applied Anthropology in 1940, and several others. In 1974, she was elected president of the American Association for the Advancement of Science under a new procedure which opened voting to the general membership for the first time. About 44,500 of the 130,000 AAAS members cast ballots; Melvin Calvin, a Nobel laureate in chemistry, was her opponent. Mead has also been invited to give a number of named scientific lectures, including the Terry Lectures at Yale University, the Joshua Mason Lectures at Birmingham University, and the Phi Beta Kappa Lecture at the AAAS. Her honorary degrees number over twenty, from institutions including Harvard, Columbia, Leeds, California at Berkeley, and the University of Delhi. In fact, a list of her honors, awards, and prominent scientific positions fills several pages of print. Ironically, journalists who rely on critics for colorful comments, implicitly rely as well on her impressive scientific credentials to justify their frequent attention to her.

In recent years, it has become more common for anthropologists to praise even Mead's attention to contemporary society. Douglas Haring defends Mead's social involvement: "Fortunately she is not inhibited by notions of the neutrality of scientists in practical affairs, and her emotional involvement in the urgency of action is based soundly on a wide range of facts sensitively comprehended." In a review, Sol Tax makes a point which Mead believes is particularly telling. In essence, Tax argues that Mead is not going out of her area of expertise when she comments on the American scene, but that this, too, is part of her field work: "It takes long to become a prophet at home. Professional anthropologists have for thirty years or more appreciated Mead's technical competence but wondered at her popular writing, lecturing, and conferencing. These may now be interpreted also as fieldwork, or 'research,' from which Mead draws significant conclusions." One of Mead's many accomplishments within anthropology, some feel, has been that she helps to bring respectability to the study of contemporary industrial societies.

Over the years a kaleidoscope of contentious comments, battles with critics, and colorful eccentricities have fallen together into an image of Margaret Mead. Her life and style are rich in image-making material of the kind journalists delight in. Her office is a convoluted series of rooms in the attic of The American Museum of Natural History, cluttered with the accumulation of fifty years' work and populated by young anthropology student-assistants. Interviewing Mead in the attic office, a *McCall's* writer let her imagination go:

To reach Dr. Mead means going through a series of checkpoints. The guard aims me through Man and Nature, and North American Forests, where enlarged bugs hunger behind plate glass; through the Biology of Invertebrates, with a side glance at good old *Balaenoptera Musculus*, the 94-foot nightmare-maker more commonly known as the Blue Whale; finally to the first floor information desk, where Gale Sondergaard's double smiles the same witch-it smile and phones upstairs to see if it's okay.

Because Hitchcock would approve, I memorize the number on my pass: 22423. . . . A quite ordinary elevator takes me, without incident, to the fifth floor and the last relay station. The final lap: a wide, hushed avenue of storage bins, yellowed signs, sinister unmarked doors. A one-way staircase, going up. The only sound my echoed steps, oddly muffled. Marienbad.

Mead's appearance and manner provide equally rich material for the journalistic imagination. The *McCall's* interviewer continues, "Vitality here. But vitality. No foxy grandma, save the shawl. Here is the cattle queen at 68 who can quite obviously ride harder, rope faster, talk better, and sit up later than anyone else. Strength, physical and intellectual. Strong, bouncy, young-girl hair. Strong white young-girl teeth. Great skin. The eyes candid, direct. Come to the point. Don't waste my time." *Life* interviewer Irene Neves wrote: "Most Americans picture her in pith helmet, pursuing primitive tribes in the jungles of the South Seas." *Time* begins: "Looking like a cross between a stern schoolmarm and an impish witch, the short (5 feet, 2 inches), broad-beamed woman in a floor-length, toga-like gown, marched onto the stage at the American Museum of Natural History last week, clutching her ever-present forked walking stick."

The images used to be of the more exotic, "pith helmet" sort, but increasingly now they emphasize her role at home: "everybody's grandmother," "mother to the world." One senses in the images, too, a consciousness that her role is a female one: "She could pass for a Midwestern housewife who had possibly taken a few judo lessons late in life."

"It is hard to tell whether I was more helped or hindered by being a woman," Mead finds. She senses subtle discrimination in her own field, and an anthropologist-critic believes there is a great deal, pointing out that Ruth Benedict, a successful and popular anthropologist, was not a full professor at Columbia University until a year before she retired. On the other hand, Mead says, "I make much better newspaper copy as a woman than I would as a man." Also, in organizations, "I am always being made president when there are two men who are fighting each other, or more; then they make me president because it usually isn't a woman who is a rival." More important, she adds, "I always did [field] work that men could not have done without risking being killed: I worked with women and children."

Mead encourages image-building about herself, in the sense that she enjoys her eccentricities. She may point out to an interviewer, for example, the clutter of mail piling up in her bedroom, or the South Seas souvenirs in her living room, characteristics that will make good copy. In essence, she helps a journalist not only with good quotations but also with good descriptive material. Says Gregory Bateson, "She caricaturizes herself." Scuttlebutt among journalists includes the notion that her dramatic, shoulder-high, forked walking stick is now decorative rather than functional. She adopted the staff, an English thumb-stick she ordered from London, after she broke an ankle in 1960 in a fall at a friend's house. "Although the ankle has long since healed," one journalist wrote, "the thumb-stick has become a Mead panache and, like Holden Caulfield's skis, it presents certain problems when she is climbing in and out of taxis, or entering elevators. At meetings, she will sometimes wave it when she wants to be recognized from the floor." The ankle had broken for the fourth time, Mead explains, "which is why I decided to use a cane and not break it any more." While serving as a legitimate precaution against another broken ankle, the staff also adds to Mead's image in descriptions and profiles of her.

In another sense, Margaret Mead is good copy precisely because she is not contrived, because she is always herself. She is the same person in her living room, in a crowded auditorium, or on a South Sea island. For psychologist Martha Wolfenstein, Mead's naturalness was one of the most striking aspects of the National Educational Television film, "Margaret Mead's New Guinea Journal." Recalling the 1967 film, Wolfenstein said:

You see Margaret coming up in an outrigger canoe, in her flowered dress, with a big broad hat — because with all the time she spent in the tropics, she really can't be in the sun much because she gets very bad sunburns. She comes up to this wharf and there are these black people there to greet her with some flowers. She and an old black woman fall into each other's arms and she's just as much herself there. It's really a remarkable testimony for someone who really feels it's one world.

And you see her talking to some old men of the tribe, who were houseboys back when she first went there, talking to them in fluent pidgin English. . . . These old men were making a big to-do because one of the young men of the tribe was erecting a big house in the village square; they thought he was getting above himself. And she was laying down the law to them, "Now, when you were boys, I remember when. . . . And you've become the head men of the tribe, but you are not going to be around forever, and who's going to take charge if you keep the young down like this." Just laying it on the line, on that island in the South Pacific.

She's just terribly much herself wherever she goes. . . . The way she comes through in this film is just like Humphrey Bogart, an American actor whose acting consists in being himself all the time, in a very vivid, forceful way. She's really a marvelous movie actress, in the best tradition of American acting, which isn't acting but being yourself.

Most reporters make some effort to come to grips with Mead's energy, productivity, enthusiasm, and vigor, pointing out that she fits more activities into a day than many people into a week or month. To put together a *Redbook* interview recently, Irene Kubota began by proposing the interview to Mead in her attic office at the museum. Still discussing when to fit in the interview, Kubota tagged along as Mead went to lecture at the YM–YWHA (to which she donates time, Kubota says, in gratitude for typing lessons her daughter received there over twenty-five years ago), then downtown to buy shoes. After waiting for Mead to return from conferences in Sweden, Greece, and Iran, Kubota spent six months catching four interviews, at Mead's office, over breakfast in Mead's apartment, at dinner at her own apartment, on a cab ride to LaGuardia and over lunch in the airport before Mead took off for another speaking engagement. Noting that Mead has been known to turn out a book in twenty-four days, give eighty talks a year all over the world, serve simultaneously on countless scientific and government committees, maintain her museum and teaching posts, and tutor innumerable anthropology students, another journalist quotes Mead as remarking that she can do things in fairly "intense doses." "The doses," the journalist concludes, "are not merely intense; they are Gargantuan."

Surprisingly little is said in print about Mead's childhood or her personal life, in spite of its journalistic possibilities. Profiles sometimes note the influence of her grandmother, and the fact that as a child of eight or nine she was encouraged by her grandmother or parents (the accounts vary) to record the patterns of speech of her younger sisters, her first foray into observation of human behavior.

Her first marriage (she had three) to a theology student, Luther Cressman, was a journalistic enigma, referred to occasionally in biography, but often omitted as if it never existed. In her autobiography, *Blackberry Winter*, Mead explains, "When I began this book I wrote to Luther to tell him about it. He and I had agreed that since ours was a student marriage, out of which neither a book nor a child had come — either of which must, of course, have been acknowledged — it was not necessary to introduce our marriage into later public records. Occasionally, an industrious journalist, going through newspaper files, has come across columns of comment, all around the nation, that greeted my decision to keep my own name when Luther and I married in 1923. Otherwise, by and large, our marriage has remained part of our young and private past."

Reporters seem embarrassed to ask personal questions of such a monumental woman. But in *Blackberry Winter,* which covers the years from her birth in 1901 through the beginning of World War II, "she never flinches from telling us," *The Saturday Review* comments, "what we've always wanted to know."

The marriage to Cressman, Mead explains in her autobiography, disintegrated after her first field trip to American Samoa, which followed difficult years while both were students. Two years later she married New Zealand psychologist Reo Fortune, whom she had met on the return boat from Samoa. She and Fortune went together on field trips to New Guinea, where in 1932 they teamed up with a British anthropologist, Gregory Bateson.

The Saturday Review picked up the human interest in the triangular situation in New Guinea and quoted *Blackberry Winter* in detail: "Cooped up together . . . we moved back and forth between analyzing ourselves and each other, as individuals, and the cultures that we knew and were studying, as anthropologists must . . . it became clear that Gregory and I were close together in temperament — presented, in fact, a male and female version of a temperamental type that was in strong contrast to the one represented by Reo. . . . The intensity of our discussions was heightened by the triangular situation. Gregory and I were falling in love, but this was kept firmly under control while all three of us tried to translate the intensity of our feelings into better and more perceptive field work. . . . Both Gregory and I felt that we were, to some extent, deviants, each within our own culture."

After returning to New York, Reo and Margaret agreed to part, Reo to marry an old girl friend and Mead to marry Bateson in Singapore on the way to an expedition in Bali. On the return voyage in 1939, Mead, who had years before been warned by doctors that it would be difficult if not impossible for her to have a child, became pregnant; advised not to travel, she remained in New York to have their daughter, Mary Catherine, while Bateson returned to England for wartime duties. Bateson and Mead separated in 1947.

The daughter and granddaughter of professional women, Mead lived the life of a 1970s-style liberated woman. "She married and shed three husbands," *The Saturday Review* points out, "without taking any of their names or following them anywhere she didn't want to go." After her daughter was born, Mead continued her career, raising the child with nurses and in a large communal family with friends in Greenwich Village. She was trying for family arrangements based on the Samoan pattern, according to Bateson. Her daughter has a Ph.D. degree in Middle Eastern Languages from Harvard University, and a daughter, Sevanne Margaret, of her own. Mary Catherine, too, has continued a career: the summer before her child was born in 1969 she went to Austria to a conference her father organized and from which she wrote *Our Own Metaphor*. In 1975, after an appointment for several years in the Department of Sociology and Anthropology at Northeastern University, she became dean of graduate studies at Damarand College in Iran, where her husband is a professor at the Iran Center for Management Studies.

The peculiar way Mead's personal life has stayed out of her public image is very different from what would have happened if she had been a politician. In fact, her colorful personal life may have been a factor in preventing her from entering politics, she explains in *Blackberry Winter*. In college, she rejected the idea of a political career on the grounds that "political success was both too short-term and too exigent." Later, however, she might have changed her mind, she says, but she felt she was barred from political life because she had been divorced twice. "In a curious way, this has both protected me and permitted me a kind of single-minded pursuit of the things I have valued, just as being a woman has protected me from having to accept administrative posts. Otherwise, with my propensity for letting life call the shots, I might easily have been diverted by the argument that it was necessary for me to play a political role. As it was, as long as I did not put myself in the position of being a political target, my private life was not a liability and, in fact, rapidly faded from most people's memories. Today, occasionally, I receive letters attacking me as a spinster without any right to discuss questions of family life."

Perhaps journalists did not delve into Mead's personal life too deeply out of a sense of fair play. Journalists have usually liked Margaret Mead, partly because she was good copy, and partly because she was always understanding of the press's needs. She has been generous with her time. When *Time* magazine was writing a profile of her in 1969, she recalls, they wanted to "jump the gun" and include in their article information about the last lecture she was giving for The American Museum of Natural History's Man and Nature Lectures. To enable *Time* to publish its article the same day as the lecture, Mead wrote the lecture ahead of time — "normally, I would have composed as I went along." As it turned out, "the man who was doing it got the measles or the mumps or something, so it didn't come out for a week later. But that's the reason I had it in that shape — because they asked for it."

In turn, journalists have generally been considerate with Margaret Mead. "On the whole," she says, "the press has always been friendly to me. They have always done what I asked them to do, virtually. If I landed somewhere and said, 'Look, this is a private visit; I don't want a lot of press,' they have respected it, because I have always tried to give them a good story if there was one. My general theory was, give them a better story than they can write themselves. They are pressed, and they're in a hurry, and if you can give them a good story, they will write that."

Reporters' descriptions of Mead have been overwhelmingly positive and complimentary — "a major force" . . . "folk heroine" . . . "as generous as she is gifted" . . . "zestfully efficient" . . . "by far the most widely acclaimed figure in her profession." "I have to confess," says Caroline Bird in *The Saturday Review*, "I would rather be Margaret Mead than me." Journalists quote from her critics, even snide remarks like Columbia anthropologist Marvin Harris's wry dig in *Time:* "The courage of one's convictions is a blessing with which Mead has been liberally endowed." But the journalists counterbalance the ripostes with compliments of their own. In particular, they defend her right to be relevant and readable.

At times newsmen have been almost protective. Mead recalls an incident in which a newsman saved her much embarrassment:

I went down to a small college in North Carolina that was pretending to be very liberal and was actually exceedingly illiberal . . . I lost my temper at them. It was a Methodist college. I said, "I'm not coming down here again until you people pay more attention to the Bible and the Constitution of the United States." And an integration person, of course, went out of there and published this.

And I got a call from a Springfield, Massachusetts, newsroom in the middle of the night, saying, "Do you know what's going out on the wires? — that you are too arrogant to go South, or something."

But they had picked it up and said, "This doesn't sound like you," and they called me at two in the morning. It followed me from one place to another, and by the time I was called by a radio station at seven, I knew what was happening.

So I really feel completely safe in the arms of the press on the whole.

Mead's good relationship with the press has of course not protected her entirely from bad moments. She expects distortion. She also expects disappointment. She noticed, for example, that after she had given a press conference at the 1972 meeting of the American Association for the Advancement of Science in Washington she found nothing printed in the *Washington Post* the next day: "That was very interesting. They gave enormous play to René Dubos. They played him as absolutely optimistic, 'man's going to bounce, everything nice with nature.' They had a lot of good pictures of him and he gave the Morrison lecture. He gave it, I think, on Thursday night and had the press conference the next morning. And I had my press conference Friday morning and didn't give the lecture until Friday night, where I had a huge crowd — and not one word in the *Washington Post*." Mead attributed the omission to timing; her speech and press conference were out of synchrony with press deadlines. But some of the reporters at the press conference declared afterward that they had been bored. Mead said nothing new, they complained. During the press conference, some of the reporters in the audience asked questions that seemed to be baiting her: "Dr. Mead, they just discovered that there's air on one of the satellites of Saturn; do you think we ought to send a colony up there, in case this earth does fall apart?" But she answered each question straightforwardly and seriously, adding humor on her own initiative. Question: "Do you think the wave of environmental concern has crested in the United States?" Mead: "I think the press will do its best to say so, because if there's anything the press loves in this country, it's waves." At times, the press conference was like a bull session, reporters and Mead arguing the environmental issue. And at times Mead was a teacher in a classroom, admonishing reporter-students who just didn't seem to get the point.

Mead was also disappointed, and outraged, by distortion she found in Winthrop Sargeant's profile of her in *The New Yorker:* "Two or three people had tried to do profiles before, and I decided that a profile in *The New Yorker* was something you only had to suffer once, like going to high school. This man was friendly, quite friendly. He didn't mean it at all — he said my room was filled with mediocre art, and he looked around and said he didn't understand why there wasn't anything from the Pacific Islands, there [in her living room] staring at all these carvings straight in the face. It was wickedly done. In his original draft there wasn't one item that was right. Then they put a researcher on it for two months who thought he was an authority and would call up and check a half sentence or something. And they said I was studying the Balinese for schizophrenia, which did incredible harm, piling up harm for ten years. That was really an abomination."

In a profile in *The New York Times Magazine*, which was generally "accurate enough," she objected to the fact that "it sounded as if I made millions. . . . Actually I put all the money I make into a research institute, anything I make over five hundred dollars." According to a colleague, a great deal of money goes to the Institute for Intercultural Studies, which she created to give small supplementary grants to anthropology students and field workers. The profile had essentially painted a picture of a successful media artist: "Early this year [1970], as a result of successive appearances on four major TV talk shows, Mead joined the American Federation of Television and Radio Artists — she is probably the only anthropologist in the country eligible for membership — thus guaranteeing herself about $320 for each show. Her kickoff speeches at meetings and conventions bring her from $500 to $1000 (or nothing at all, if she decides the sponsoring organization can't afford the money); she draws . . . a pension from the American Museum of Natural History, a stipend from *Redbook,* for which, in collaboration with Dr. Rhoda Metraux, she writes a monthly article, and royalties from numerous books that date back to 1928, when she published *Coming of Age in Samoa,* which is still available in four paperback editions that sell at the merry clip of 100,000 copies a year."

Mead has rarely had the experience of submitting an article or book that was turned down by a publisher. For one thing, most of her writing is done in response to a publisher's request. "I suppose I have submitted half a dozen articles in my life to anything spontaneously." Occasionally, she has tried submitting an article, as an experiment. "I tested one article out — I guess it was in the early 1930s — called 'Anaesthesia for the Baby's Sake,' pointing out that the societies that thought about the welfare of the parents versus the child were bad societies and those who thought about the child were good societies, and discussing the use of anaesthesia in childbirth. It was turned down by every magazine it was submitted to. I let my agent try it. They were all edited by men, and they all *knew* that women couldn't stand the pain of childbirth. That was an experiment, to see what would happen."

Once in a while, she says, she has also submitted an article "for somebody else's sake in some way." On one occasion, she started a joint article with a photographer who had taken pictures of families watching television, contrasted with pictures of the same families reading. "I wrote an article to show why people who read a lot hate television, what the problem is. *TV Guide* turned it down *flat.* They should have published it. They were fools not to."

In all her ups and downs with journalists, Mead points out that she has one advantage over most scientists: when she takes field trips, as she does nearly every year, she gets vacations from the press. "If you are a laboratory person and have a big lot of administrative responsibilities, you know, somebody comes scratching at that door. It's very hard to find me when I have gone to New Guinea. I just say, 'Good-bye, I've gone.'"

The Popular Connection

For the survival of our civilization and our species, as well as for what philosophers since Socrates have recognized as the most profound of human pleasures, I believe there is an urgent need for better, serious and more widespread popularization and understanding of science.

— CARL SAGAN

In the constellation of visible scientists, the newest star is Cornell astronomer Carl Sagan. Set a little apart from the other stars, he is in one sense a science popularizer. He is a cosmic consciousness-raiser, stirring intellectual, cultural, and financial support for space exploration in particular and the scientific way of thinking in general.

But Sagan's traditional science-is-wonderful message is transmitted by the very untraditional paperback-and television media. It is a new kind of popularization, a crossbreed of old ends and new means. And as such, it points up changes in the communication of science. Sagan's visibility is a product of the 1970s, a process occurring now, and a prediction for the future.

A visible popularizer is an unlikely phenomenon at a time when the public takes a jaundiced view of technological wonders and a jaded view of men-on-the-moon. What makes the difference for Sagan is talent and timing: he has the makings of a television celebrity and a topic that resonates with public interest. Extraterrestrial life, as he has observed many times, is an idea whose time has come; and Sagan, as *New Scientist* has added, is a "man whose time has come."

Popularization works for Sagan because Sagan is a space expert, space is a television topic, and television is Sagan's medium. The convergence is a coincidence of the 1960s, an aftermath of *Sputnik*, when space exploration became politically expedient, technically feasible, scientifically legitimate, publicly relevant, and journalistically fashionable. Riding the crest of the new interest, Sagan spearheaded both scientific and journalistic developments. "As I often say, if I were born fifty years earlier, I could speculate all I want but I couldn't do a thing; born fifty years later, it would all have been done." Young (early forties now, lively dark eyes, dark fashionably long hair), ebullient, engaging, and very articulate, he is as natural to television as his topic.

The result is popularization because Sagan uses the media to share his vision of the cosmos and the scientific way of reaching it — astronomy not astrology, space probes not UFOs, logical theory not Velikovsky. ("The popularity of astrology," he says, "is an unhappy commentary on the lack of toughmindedness and the dearth of open and critical thinking in our society.") He is a master at the art of the mini-lecture, a quick and painless injection of education that can be administered between quips on the Johnny Carson show, between commercials on a documentary, over coffee with science writers:

If the eons of earth's lifetime were compressed into the span of a single year, the origin of life would have occurred at the end of January; the colonialization of the land in November; the flourishing of the dinosaurs on December 15; the evolution of the mammals on Christmas Day; the development of the first primates near dawn on December 31; and the origin of man at 8 P.M. on New Year's Eve. Recorded history would occupy the last 30 seconds of the last day of such a year.

The same mini-lectures also work between pictures of a paperback, expanding Sagan's audience to the print-bound public. *The Cosmic Connection*, on its way to selling a million copies, is a cafeteria of adventure stories, science fiction scenarios, and miniature courses in astronomy and biology — thirty-nine chapters averaging less than seven pages each. One can flip chapters like channels and find Sagan tangling with the CIA in Berkeley; being propositioned by a dolphin in a warm Virgin Islands pool; caucusing with Arthur C. Clarke and Stanley Kubrick about how to rescue the deadended script of the film *2001;* viewing (the first person for the first time) the Martian moon Phobos on Cal Tech's *Mariner 9* monitors; taking imaginative trips through black holes to "elsewhen"; sobered by the sophisticated cosmic perspective of Mill Valley first-graders.

If *The Cosmic Connection* (1973) is a documentary, *Other Worlds* (1975) is the "Tonight" show, a collage of pictures, poetry, cartoons, quips, and mini-lectures, produced (arranged), as was *The Cosmic Connection,* by Jerome Agel, along the lines of Agel's earlier book with Marshall McLuhan and Quentin Fiore, *The Medium Is the Massage.* Sagan says, "People told me that *Cosmic Connection* was okay but it was too hard for a lot of people, which I found very astonishing. But maybe it was true, so I thought I'd do a little book that wouldn't be too hard for anybody . . ."

Sagan now has a four-book contract with Random House, from which *The Dragons of Eden,* based on a series of lectures on the evolution of human intelligence, is being published in 1977. Sagan is also working with director Francis Ford Coppolla on a six- or seven-hour NBC television special to run two or three nights on prime time, a fictional scenario of man's first radio contact with life in outer space.

Mini-lectures are also the medium for hassled science reporters, for whom Sagan's office at Cornell has become Information Central, just as it is for exobiologists themselves. Along the walls on the way to Sagan's office on the third floor of the modernistic Space Sciences Building, interviewers encounter a crash course in astronomy. In 1975 the lobby featured three-dimensional pictures of the moon's surface, with special glasses on chains, like bank pens, for viewing the rugged surface in 3-D. The rest of the information is all there in Sagan's office, piled high in stacks of correspondence, review copies of books, drafts from colleagues. To spice up a story, a writer can always mention Sagan's bright orange Porsche with its California vanity license plate "PHOBOS" (a moon of Mars); his involvement in the latest *Mariner* or moon flight; his casual clothes (a lot of turtlenecks); his tendency to wrinkle his nose; his trouble with colleagues who resent his cavalier approach to detail. To bolster the credibility of a scientist who seems so young, speculative, irreverent, humorous, and altogether human, a writer can toss in a mention of Sagan's many honors, awards, professional offices, and committee positions, both as a popularizer and as a scholar. "He's a real showman," said *New York Times* science editor Walter Sullivan, early in Sagan's career. "He always makes good copy when he talks."

Sagan's starting point is his own intense enthusiasm, a fulfillment of dreams that began well before there were scientists or *Sputniks* to provide encouragement. "I sure didn't pick this subject because the public was interested in it. I got into the subject when everybody told me it was ridiculous." Sagan decided to be an astronomer before World War II — at the age of five. Legend has it (correctly) that, growing up in a poor neighborhood in Brooklyn, Sagan paid special attention to the stars. Somehow they did not seem to belong to the alleys, stoops, and empty lots he knew as Brooklyn. But when he asked grown-ups about the stars, their answer did not satisfy him: "They're lights in the sky, kid." Just at that time, having learned to read very early, Sagan was given a "cherished possession," a library card. His first request at the library was for a book on stars. The diminutive five-year-old was duly offered picture stories about Alan Ladd or Veronica Lake, but after some negotiation he was finally given a children's book on astronomical stars. He sat down right in the library to read it: "It was the sense of scale which first grabbed me. It was awesome. In a sense, it was religious. The universe was much vaster and more awesome than I had realized." A few years later he extended his fascination to the question of life on other planets. "I discovered that we were on something called a planet, and it went around the sun; we were small and didn't give off our own light. If something like that were going around stars that you can barely see, certainly you couldn't see those other planets. Other planets, worlds — suddenly I was just very excited. . . . From that early epiphany I never expected to do anything except to pursue this personal sort of passion."

In Sagan's world, however, it was one thing to dabble in astronomy and another to earn a living. In the opening lines of

The Cosmic Connection, Sagan recalls that when he was twelve his grandfather asked him what he wanted to be when he grew up. "An astronomer," Sagan answered. "Yes," said his grandfather, "but how will you make a living?" Sagan assumed he would have to support himself with "some absolutely crummy, dreadful, awful job that I was temperamentally unsuited to . . . like door-to-door salesman," while astronomy was consigned to evenings and weekends. "Understand," he points out, "that I not only didn't *know* any scientists; I didn't even know that there *were* any scientists." Finally, in his sophomore year of high school, Sagan's biology teacher told him he was pretty sure that Harvard paid astronomer Harlow Shapley a salary, and that he wasn't a door-to-door salesman part-time. "That was a very glorious day." Even so, Sagan remained a bit of a closet astronomer until after *Sputnik*. "Uncles of girl friends, at dinner, would say, 'Fly to the Moon? Buck Rogers stuff?' And then they would advise the girl to leave me alone: I was obviously crazy. There was a period pre-*Sputnik* when I sort of kept to myself what I was really interested in."

Problems were postponed at college: Sagan chose the University of Chicago because its promotional brochure emphasized education, and he got just that. The University of Chicago had a compulsory liberal arts curriculum. "Zero electives. The faculty decides a set of courses which everybody who wants to be educated has to go through, and at the end of that you get an A.B., which serves you for absolutely nothing afterwards, except you are declared educated (and to a surprising degree you finally were)." Impressed with the plan, Sagan even audited courses that his high scores on advanced placement exams had allowed him to skip. Without that broad education, he reflects, "I could have remained immensely narrow." It also provided training for later popular speaking and writing.

The next hitch was that, while astronomy might be a legitimate career, extraterrestrial life was, even for astronomers, too far out. In graduate school in astronomy at Chicago, there was no way for Sagan to get formal training in molecular biology; it was simply not available. In fact, as Sagan enjoys recalling, there were no molecular biologists at Chicago to sit on his thesis committee — with one exception: Kimball C. Atwood, a professor of obstetrics and gynecology. "So I believe I am the only astronomer in the history of the world to have had on his thesis committee a professor of obstetrics and gynecology."

In these pre-*Sputnik* years, Sagan kept up his unpropitious interest in extraterrestrial life largely through the encouragement of two Nobel laureate geneticists outside of Chicago: Herman J. Muller and Joshua Lederberg. "If not for Muller and Lederberg," says Sagan, "I would have had a very tough time maintaining my resolve." He met both by chance. "I'm home spring vacation of my first year at college. There I have my sneakers on, basketball under my arm, walking out the door, and my mother says to me that she has a friend in town whose nephew is a scientist. Wouldn't I talk to him? And I said, 'Oh, Mother, there's all kinds of scientists, and I don't want to go.' And she said, 'Look, he's probably an interesting guy; do it for me.' I said, 'All right, I'll see him later.' And I went out to play basketball, came back, went to see this guy. His name was Seymour Abrahamson [then a graduate student at Indiana University]. He's now a professor of genetics at University of Wisconsin. And it was more interesting than I had thought. I didn't know much about genetics, but I certainly knew it was relevant to worrying about life elsewhere in the universe. So I got to talking with him about the molecular basis of life and so on. Then when I went back to school, I wrote him a letter, just on some other points of the conversation. He showed the letter to Muller."

Muller invited Sagan to work in his laboratory at Indiana University that summer (1952), and, although Sagan spent most of his time looking at fruit flies under a microscope, "separating guys with bristles from guys without bristles, guys with big red eyes from guys with little white beady eyes," Sagan found the contact with Muller invaluable. Having no formal biology training, Sagan learned as an apprentice, first with Muller, then Lederberg.

Sagan happened to meet Lederberg because he could not tolerate life in Williams Bay, Wisconsin, where Chicago's Yerkes Observatory is located and astronomy graduate students are based. For one thing, Sagan was arrested in Williams Bay for campaigning for the Democratic Party. "It was called Dollars for Democrats Day, and the idea was, at least as I understood it then, that since the Democrats had virtually no money even for campaigning, this was a way of giving them some money so we could hear both sides of the argument. My pitch wasn't even that they were right; just to let them talk. So I spent all morning going door to door. And I got the most amazing responses: 'The *what* party?' or 'Shh! the master will hear!' or 'Wait right here, young fellow, and I'll get my shotgun.' Finally I was arrested by the sheriff, who had had innumerable complaints, on the grounds of peddling without a license. They figured I was peddling receipts at a dollar each. And I was remanded to the custody of the observatory director, who I don't think understood anything about it, but just said to me, 'Be a good boy.' " At Williams Bay, there were also anti-Semitic incidents, Sagan says "the sort of things I never had before or after in my whole life." So he moved to Madison, and commuted to work from there. And in Madison, through mutual friends, he met Lederberg, who was at the time professor of genetics at the University of Wisconsin, and just getting interested in space biology. "In fact, he was trying to learn about planets, so I was in this very silly position of tutoring *him*." They struck up a friendship; Sagan later spent a semester with Lederberg, who had moved to Stanford, and they continue to exchange ideas. In the meantime Lederberg coined the term "exobiology" for the study of extraterrestrial life.

At about the time Sagan met Lederberg, *Sputnik* was launched, and Sagan graduated into the space age. As government and scientists rallied to the challenge, word quickly got around that Sagan was one of few available experts on the subject of extraterrestrial life. Before his Ph.D. was in hand, he was asked to serve on study panels and committees for the august National Academy of Sciences and the newly formed National Aeronautics and Space Administration (NASA). Indicative of the changing academic mood toward exobiology were two conferences on the subject held in 1961 and 1971. In 1961, a dozen or so scientists gathered at Green Bank, West Virginia, to discuss extraterrestrial life. "Nothing much came of that meeting,"

Sagan recalls in *New York* magazine. "We passed out Order of the Dolphin lapel pins and got smashed on champagne," celebrating the announcement that one of the participants, Melvin Calvin, had just received the Nobel Prize. In 1971, over fifty scientists met in Soviet Armenia, a conference that Sagan concluded "made the subject of communication with extraterrestrial intelligence scientifically respectable." Sagan chaired the American delegation and edited the proceedings, which were published as *Communication with Extraterrestrial Intelligence (CETI)*.

Even ignoring the passions and fortunes of exobiology, Sagan simply did not have the makings of an obscure scholar. He was, as his high school classmates had voted, "most likely to succeed." While a graduate student he worked on at least two unusual theories that made waves in the astronomy community. First, he speculated that color variations observed on the planet Mars, which had been thought to indicate the presence of seasonal changes in plant life, were instead the result of wind storms causing shifts in layers of dust on the planet. Sagan developed the idea further while on the faculty at Harvard in the 1960s, and his interpretation was dramatically corroborated by information sent back by *Mariner 9* in 1971 and 1972. Sagan, while a graduate student, also proposed that the dense atmosphere of Venus, composed of carbon dioxide and water vapor, held in the sun's heat, like a greenhouse, producing extremely high surface temperatures on the planet — Venus was "hell." This idea, too, was confirmed by data, this time from the Soviet space vehicle *Venera IV*, which entered Venus's atmosphere in 1967. Says British journalist Ian Ridpath, "Sagan has . . . made a habit of extrapolating adventurously from limited data — and being infuriatingly right." Sagan is also known in the scientific community for experiments, following on the work of Stanley Miller, in which he simulated the atmosphere of Jupiter and primitive earth, determining that organic molecules could be produced in such environments. Indicative of his scientific success, Sagan has climbed the academic ladder rapidly since getting his Ph.D. at the age of twenty-six: post-doctoral fellow at

Berkeley, visiting assistant professor at Stanford (where he worked with Lederberg), assistant professor at Harvard, professor and director of the Laboratory for Planetary Studies at Cornell University's Center for Radiophysics and Space Research.

As space exploration came to dominate science coverage on television, word also got around that Sagan was a good talker. "It wasn't so much that they cared what I said, just that I seemed to say it in a way that didn't put people to sleep." After warm-up appearances on documentaries, the "Today" show ("There we were at launch pad 43, sitting on chairs out there, and the wind was blowing, and Frank McGee was asking good questions."), and a Dick Cavett special on UFOs, Sagan caught the attention of the instant celebrity-maker, Johnny Carson. "He had seen me on the Cavett show," Sagan recalls, "and said, 'I want that guy.'" Bolstering Sagan's chances, *The Cosmic Connection* was just about to come out, and Doubleday did the usual urging to get its new author on the book-plug section of the show. Also, Sagan had a friend, an actor in Hollywood, who in turn had a friend working on the "Tonight" show who (without telling Sagan) added to the clamor for Sagan.

Sagan's first performance in 1972 was a show-stopper. "He went out," says writer Stuart Baur, "as a young astrophysicist with a solid reputation in his field and flew back to Cornell as that rarest of creatures — an effective media-scientist." Baur devotes two full pages of *New York* magazine to describing how Sagan "roused ten million or so Americans out of their hypnagogic stupor"; Baur begins:

He had to wait through an opening monologue; badinage with Ed McMahon; a visit from Karnak; an imitation Baez; an appearance of the Mighty Carson Art Players; a Vegas comedian glistening with flop-sweat; a pair of harmonica-playing hillbilly twins; and a gifted crow in a cage who could talk, but wouldn't, not that night, not for all the birdseed in the world.

By this time Carson was a carnival of facial tics and fidgets. Almost as an afterthought, during the book-plug segment, at a quarter to one, he invited Sagan out. . . .

After the hour and a quarter of careerist drivel from the headliners on the couch, Sagan had a cleansing effect. He burst one balloon — the credibility of UFO reports; launched another one — the possible profusion of extraterrestrial life. . . .

Three weeks later, after the 12:30 network break, he reappeared on the "Tonight" show. This time Carson gave him carte blanche, and Sagan launched into a cosmological crash course for adults. It was one of the great reckless solos of late-night television. It presumed a lot. It hit the bull's-eye. . . .

When Sagan finished and settled back into the eye of the hush he had generated, one was willing to bet that if a million teen-agers had been watching, at least a hundred thousand vowed on the spot to become full-time astronomers like him.

Counting reruns, Sagan can be found on Carson's late-night program about half a dozen times a year (he does the show two or three times a year). "I didn't intend to be a permanent fixture. When I was [first] asked, I thought, 'Okay, I'm going to go on the show once.' They asked me back, and I figured that was a formality, but then they really asked me back, so 'Okay, I'll go back another time.' It's only in the last year that I realized that I really am there a fair bit."

Sagan's prespace-age contacts with the press had been inauspicious. His first year of graduate school, 1956, happened to be the year of a big Mars opposition: the planet was the closest it had been to earth in thirty-two years. Sagan went with his advisor, Gerard Kuiper, to McDonald Observatory near Fort Davis, Texas, "to kind of learn what Mars was like through a telescope." Both Kuiper and his twenty-two-year-old student reported on their observations at the next meeting of the American Association for the Advancement of Science, December 1956, in New York City. The *New York Times* reported the two talks in one article but attributed some of Kuiper's remarks to Sagan and vice versa. "Since we disagreed, me and my major professor, it was a source of some embarrassment for me. Kuiper never said a word about it; for all I know he never knew or it didn't bother him. But for me it was dreadful. I was immensely flattered that the *New York Times* would quote [my talk], and then amazed that they would get it wrong — they were two papers in two different sessions. How could they have confused it? And if they get that wrong, what do they do about politics, which is a lot mushier? . . . And the final thing was, I came back through [Chicago] and a faculty member there said, 'I've been following your career in the *New York Times*.' He strongly conveyed the sense that scientists are not supposed to be quoted in the newspapers. . . . I had the sense of having done something bad by being quoted. . . . You know, the life of a graduate student, especially then, was very tenuous. You're very much at the mercy of faculty, and they can, at the very least, make your stay unhappy or at least longer. That got me thinking a little bit, and I thought I would leave that sort of thing alone, for a while at least. I was sort of shocked by that, and felt I had been burned somehow. Nobody had ever explained to me that you get burned by giving a scientific paper at AAAS, but somehow it had happened."

Five years later, Sagan's second press encounter was also unnerving. Speaking at the Radiation Research Society, "an extremely obscure society," in San Francisco, Sagan stated, on the basis of his dissertation, that Jupiter was a planet where organic molecules might be forming today, because conditions were similar to those that produced life on earth. The society asked him to do a press conference, recognizing undoubtedly that here was a speaker who might attract reporters. "Having had that experience with the *New York Times* a few years before, I was worried about getting it wrong. So I said to the reporters, 'Look, I'm talking about organic molecules, not life. I'm not saying that there's life on Jupiter. I'd be very unhappy if any of you wrote that I said "life on Jupiter." ' The next day a San Francisco paper reported: 'Life on Jupiter, scientist says.' "

A television man at heart, Sagan has remained uncomfortable about newspaper and magazine coverage. "Sometimes you feel like you have no control; it doesn't matter what you say: they'll say whatever they want — very frustrating." He counters inaccuracy with attempts to educate reporters. "There aren't all that many science writers, so I've come to know who's who; I can remember their names, faces. Sometimes we have coffee together. So now if I say, 'Hey, look, what I mean is this not that,' I have the sense that maybe they pay a little more attention, or maybe they don't want me to be mad at them next time." But as Sagan's visibility has increased, so has the number of inexperienced reporters, who are still inaccurate. And so Sagan continues to worry whenever he gives a public talk.

Until recently Sagan took publicity as it came. "I never decided what I was going to do; I just did things as they came along." He recognized his visibility slowly — in fact, the thought did not crystallize until he read about the dissertation on which this book is based. In 1971, however, *Time* magazine did a feature article, "Is There Life on Mars — or Beyond?" in which Sagan was the only astronomer pictured, the author of the opening quote, and the scientist dubbed "exobiology's most energetic and articulate spokesman." Says Sagan, "They said some nice things about me, and that was a large audience, so I suddenly had the idea that I was visible." Three years later, in 1974, *Time* included Sagan in its list of "200 Rising American Leaders," and *Saturday Review* named him one of thirty-two "leading professionals."

"When people started recognizing me on the street I was astonished. I went to get a pair of shoes at a shoe store here in Ithaca, and in paying for them, I told the clerk (who I never saw before in my life) that I had bought the same kind there before and had worn them every day for a year and had liked them fine. He said, 'Really? Did you even wear them on the Johnny Carson show?' "

Another sign of Sagan's visibility: he is beginning to go through the painful process of setting priorities among the exhausting demands on his time — teaching and research (first priority), time for the family (his wife and precocious young son, plus two older sons by an earlier marriage), administrative duties (advisory committees and professional societies), public activities (how much? which? when?). "Once I had a conversation with Lederberg when I was just beginning to realize that I was getting a lot of invitations; I did a calculation and I said, 'You know, if I went to every meeting I'm asked to, I could spend all my time going to meetings.' Lederberg said, 'That's nothing. If I spent my time just *answering* all the invitations to go to meetings, I could spend all my time doing nothing else.' "

Visibility has been a by-product of Sagan's intense campaign to vindicate his vision of extraterrestrial life, to garner financial support for science, and to bring laymen into the fold. One senses that while he recognizes the political implications (he once quit an Air Force Scientific Advisory Board panel because he foresaw the possibility of its involvement in the early Vietnam War), he is motivated chiefly by the scientific considerations. Space probes are a direct fulfillment of his daring curiosity. Social justifications for the immense outlay of funds (space exploration could keep the military too busy for wars, and besides the cost is less than most missile project overruns) are secondary. Intellectual and cultural justifications are mind boggling. It is a matter of emphasis.

Increasingly, Sagan is also in the exhausting position of justifying the scientific enterprise as a whole, to a fundamentally skeptical public. He does it reluctantly, out of fear no one else will. "There are an awful lot of antiscientific enterprises around, which are on the edge of science. I'm thinking of things like UFOs, pyramidology, the emotional life of plants, Uri Geller, prognosticating the future, psychic surgery . . . all of which are antiscientific, marked by a complete absence of critical thinking. I can't imagine a clearer example of public credulousness than presidential politics for the past five or six years." The scientific way of thinking must, then, be promulgated, for the very health of the nation.

Popularization put Sagan in the visibility game, and it keeps him there in spite of the difficulties. "Except for what I'm about to say," he reflected recently, "I would think the negatives outweigh the positives by a large factor. But I get kids writing to me and saying: 'I never realized what an exciting thing it is to be interested in astronomy; I'm very bright and I get all these good grades, and I never knew about astronomy. Now I'm going to be an astronomer.' Now suppose one really good person becomes an astronomer because of such early stimulation. One really good person in astronomy can make a huge difference. It can be the most important thing I do, much more important than any of my research, to turn some exceptionally bright little kid on to science.

"When I was in high school, I wrote letters to a bunch of astronomers, and some of them actually answered me. I got answers from Mt. Wilson and Palomar Observatory — somebody from there would write to me, a little kid in Rahway, New Jersey! That was a big factor, the sense that they weren't immensely distant, only a little distant. And that seems to me to be the principal social function of visible scientists."

At center stage in the space age, Sagan uses rocket technology as well as media technology to communicate the cosmic perspective. For the past two decades earth people have been shooting off rockets to the moon, planets, and stars, "like a dandelion gone to seed," Sagan points out. *Pioneer 10* is on its way to the stars with a plaque aboard designed by Sagan and his wife, an engraved "greeting card," depicting man and his solar system, an invitation to star folk, a rocket-propelled bottle cast into the cosmic sea. Radio messages are also being beamed to outer space, all part of the same effort.

The objective is no less than a cultural revolution, a space-expanded view of man's fragile existence in tenuous living quarters. It would take thousands of years or more to get a return message from a distant civilization; Sagan is looking for responses from this one.

Hyping Science

Hype:
  v.t., to stimulate artificially as if with hypodermic injection;
  n., something that stimulates sales or interest, a publicity
  stunt, campaign . . .

  — *The World Book Dictionary*
  (1974 edition)

When Stanford University engineer William Shockley, known
for unpopular views on the heritability of IQ and the inferior
IQ of blacks, schedules a new genetics speech, he makes a few
strategic telephone calls. He calls a sympathetic (typically conservative and/or Southern) newspaper editor and mentions the
forthcoming lecture. The editor has no reporters he can send to
such a distant event, and agrees he should have Associated Press
cover it for him. The newsman calls Associated Press, and
Shockley is practically assured of nationwide wire service coverage of his speech.

Shockley is an extreme, and extremely unpopular, example,
but he is hardly alone in his attempts to manipulate the press.
Cultivating the press in order to promote a person or policy is
called "operating" among reporters, and "science operators" are
among the best in the business. One scientist has perfected a
technique for an "instant press conference." The last afternoon
of a scientific conference, when the science reporters are sitting
around wondering what to write, he appears in the press room.
Someone called him to come for an interview, he announces, but
he can't remember the name. All the reporters light on him, an
answer to their writing quandaries, and he has in essence an im-
promptu press conference. At major scientific conferences, the
press room usually has a guard at the door to discourage intrud-

ers. To get by the guard, another scientist says, he tells the guard he has to see a certain reporter, and gives the name of one he knows is covering the conference.

Science operating is public relations practiced by an individual instead of an institution. The object is to sell science and scientists to the press, in a system where almost no one gets attention without seeking it. The techniques are the time-honored traditions of press relations: take a reporter to lunch, stage a press conference, find a gimmick. The basic strategy is to cultivate receptive reporters, do favors for them, and ask favors in return. The willing scientist is in a good bargaining position, since science writers need friendly scientists to alert them to developing stories, to explain technical points, and to put them in touch with shyer and more recalcitrant colleagues. In exchange, the reporter can suggest ways for the scientist to further his publicity and introduce him to other sympathetic reporters. A Harvard University professor, dedicated to ending the Indochina War, developed a relationship with a particular reporter in Cambridge whom he helped with background information and referrals to other scientists. In return, the professor says, he called the reporter when he needed help getting press coverage for a new Indochina story. A midwestern professor says he always makes a point of cultivating a few reporters wherever he travels, offering them drinks, giving them a new story. H. Peter Metzger, Colorado's maverick and highly effective environmental muckraker, explains quite frankly how he was able to succeed: "Above all, I curried favor."

A good operator has a mental notebook of sympathetic, capable reporters and knows almost as much about who's who in the science writing profession as the writers do. He also learns the strengths and weaknesses of the media. Metzger, finding himself in a "media backwater" where his environmental exposés were repeatedly ignored, "went over the heads" of the Denver press to the national media. The higher he went, he discovered, the easier it was to get a controversial story published. And once his story had national coverage, it had enough clout to get into the local press.

Also in any publicist's repertoire is the art of the "pseudoevent," the technique of making something that is considered not "news" become a "news event." Daniel Boorstin, who coined the term, described a pseudoevent this way: "It is not spontaneous, but comes about because someone has planned, planted, or incited it. Typically, it is not a train wreck or an earthquake, but an interview. It is planted primarily (not always exclusively) for the immediate purpose of being reported or reproduced. Therefore, its occurrence is arranged for the convenience of the reporting or reproducing media." A news conference is a typical pseudoevent. What is presented at the conference becomes a timely event; the conference is the "news peg" on which the reporter hangs the information the conference spokesmen want conveyed.

A press conference of course works only when handled well. One effective science publicist explains that he makes a practice of preparing for a press conference with a fresh, new story, documented by pictures the reporter can use. He uses short sentences and key words that can stand on their own as quotations. He emphasizes the prestige of the event's participants; several Nobel laureates, for example, are usually among the signers of a political petition. And, whenever possible, the story is one that is also being reported at a regular scientific conference seminar, thus increasing its credibility. The scientist is uncomfortable about using academic reputation this way, using the very elitism he finds wrong with our society. But his alternative is the frustration of some of the younger, more idealistic scientists, active, for example, in Science for the People (formerly SESPA). At the 1972 American Association for the Advancement of Science meeting in Washington, the contrast was evident. At a press conference held jointly by SESPA and the more "establishment" anti-Vietnam scientists, the SESPA speakers were dull, polemical, deliberately unaccommodating; the reporters settled back, put their pencils down, and obviously tuned out. The organizers had interspersed the SESPA speakers among more dramatic media-oriented spokesmen to save the day. Later the organizers got an invaluable assist from a CBS reporter who told them, "If you decide to take your petition to the White House, let us know." It hadn't occurred to the group to deliver the petition to the White House personally. They made the trip, accompanied by television cameras, and were turned away from a silent, unresponding White House, a dramatic scene shown the same evening on national CBS news.

Operators come with varying motives and meet with varying success. If a scientist seems to be pushing a worthy cause, one with which the reporter can identify, a reporter listens, until he becomes cloyed or annoyed. If the publicity seeker is more transparently after personal fame to boost his career, the reporter censors more heavily. Discussing the "bad guys" among operators, *New York Times* science editor Walter Sullivan has remarked, "We know some of these operators pretty well. We like to think they can't flimflam us, though they probably do once in a while. There are a few commercial entrepreneurs who flimflam us once in a while, too, but we do our best to avoid it."

Most heavily criticized by the press have been some heart surgeons, whose publicity seeking, many reporters believe, has been for personal aggrandizement. Physicians and press alike have deplored the "circus-like atmosphere" and the "suture-by-suture accounts" of Christian Barnard's heart transplants. Some of the Houston heart experiments of Michael De Bakey have been dubbed surgical spectaculars, more like a big space shot than an operation.

For unpopular scientists whose views are criticized or ignored by fellow scientists, the media may represent a chance for a more sympathetic forum. Reporters speculate that scientists who are in the minority, like William Shockley, turn from the disapproving scientific community to a more receptive general audience. Some scientists may also use personal visibility in the press, thinking it will enhance the acceptance of their controversial ideas in the scientific community.

In still other cases, public recognition becomes a kind of back staircase to scientific success — not the usual way of arriving, but a pragmatic alternative for the scientist with flair and finesse. It is difficult to establish cases where disreputable scientists have actually fashioned a successful career by misusing the media, but many reporters and scientists fear that they have. Philip Abelson, editor of *Science*, speculates that it is becoming easier for science operators to get away with false claims in the press. "The publicity-seeker can get his story direct to reporters. . . . Using shrewd tactics an 'operator' can establish himself as a newsworthy person. This can open the road to research grants and even academic advancement. Only a few scientific specialists will suspect that he is a phony, and they will have no practical mechanism for penalizing him."

Obviously, too, a number of scientists suspect, and they may well be right, that visibility means research money. A scientist whose name is known to the money-givers stands out from the flood of other applicants for funding. One of the scientists involved in the race for discovery of a new elementary particle, the Omega-minus, lamented the fact that the discovery leaked out slowly and inaccurately ahead of the scheduled publication date and thereby lost its publicity — and money — value: "It was a big mess. It killed our publicity. We would have got first page *New York Times* Sunday, which is a very good thing to get. After all, where do we get national money? — from Congress. It makes a great deal of difference if we get first page on *The New York Times*, as most Congressmen read the front page. It's a factor and we can't deny it, so it matters an awful lot to us." For the scientist seeking private contributions rather than government funding, the connection between visibility and money is even more direct. A colleague has speculated that Denton Cooley reversed his earlier shyness toward the press when he broke with fellow heart surgeon Michael De Bakey and needed donations for his own Texas Heart Institute.

The press has mixed feelings about science operaters, as indeed the public would if it were aware of their existence. In a 1972 study, sixteen of seventeen science reporters felt they had dealt with operators, although fourteen believed the public was not aware of them. Some of these reporters also mentioned that operators receive unfairly extensive and unbalanced treatment in the press, while the public is none the wiser.

Reporters gossip about science operators behind their backs, but says a veteran science writer, "We are very kind to them." The need for eager, coherent, cooperative scientists remains. And reporters recognize that there is a gamesmanship to capturing press attention. Properly done (a good motive and a good story), operating is rewarded with publicity. Dramatically done, it is rewarded with visibility.

Many visible scientists were viewed as operators when they were on their way up. Barry Commoner, Carl Sagan, Paul Ehrlich, Albert Sabin, and James Watson were all branded as operators in the 1972 study of science reporters, usually with the afterthought that their operating days were in the past. Visible scientists also suffer from guilt-by-association with operators. There is a thin line between being cooperative and courting the press, and whether most of the visible scientists are guilty of publicity-seeking depends on where the line is drawn.

Once elevated to visibility, however, scientists usually shed the "operator" label, although they continue of course to use many "operating" tools and a bit of both Madison Avenue and Hollywood in their efforts to sell ideas and information. When a scientist has more desire to see the reporter than the reporter has to see him, it is called operating; when the reporter has more desire to see the scientist, it is called fame.

Real "operating" remains among the we-try-harder scientists seeking to emulate the highly visible scientists. Because the operators' enthusiasm seems pushy rather than personable, or their topic seems trivial rather than relevant, or their pronouncements are tedious rather than good copy, they continue to make an extra effort to get themselves heard.

William Shockley

That operating usually occurs among scientists with limited visibility, who have to "try harder," is exemplified by the case of William Shockley. It seems surprising at first that Shockley has any difficulty commanding publicity: one could not ask for a "hotter" topic. His views on intelligence and race engender stark hatred in blacks, hysterical animosity in white liberals, and fierce loyalty in white racists. Although he claims that the racial aspects are not central to his dysgenic theories, he has allowed himself to be diverted more and more to discussion of his belief in the inferior intelligence of the black race.

Except for his controversiality, however, Shockley has little to recommend him to the press. Making unpopular and unprecedented allegations, he has needed strong credentials. His Nobel Prize in physics, shared in 1956 with John Bardeen and Walter Brattain for development of the transistor, does not carry enough weight, many feel, when his subject matter is genetics, sociology, and education. (To counter the accusation, Shockley stresses his war experiences in operations research.) The fact that he is a known expert on transistors and considered a brilliant mind in applied physics may even hurt his credibility as he switches to genetics. And as the issue has expanded, to include Arthur Jensen, Richard Herrnstein, and others with more relevant credentials, Shockley has come to seem one of the lesser, and less credible, figures in the larger picture.

Shockley has hurt his credibility also by displaying personality peculiarities which make it difficult for some reporters to take him seriously. When an eminent Stanford University psychiatrist in 1973 remarked casually to a West Coast reporter that Shockley showed symptoms of a classic case of paranoia, the reporter found he was unable to dismiss the idea out of hand. Idiosyncrasies like his ever-present tape recorder contribute to the image of mental oddity. Shockley tapes all interviews and many seemingly inconsequential conversations; his telephone has the dramatic periodic "beep" indicating a tape recorder is installed; and all tape recordings are copied and put in safekeeping, following an elaborate indexing system. Conversations with different people on the same cassette are separated by countdowns; for example, the tape of a telephone conversation for this book wound up with Shockley announcing: "Goodell three, Goodell two, Goodell one, Goodell zero. The time is now ten minutes to seven on Tuesday, the twentieth. Goodell zero." A gadget man, Shockley has for many years tape-recorded his physics classes; he also routinely takes Polaroid pictures of the blackboard and of students to refresh his memory between lectures. He feels the tape recorder is the single most important application of the transistor.

Also hurting Shockley with the press has been a writing and speaking style that is often turgid and equivocal, reflecting his urgency and intensity as he tries to fit in all he has to say. In a generally sympathetic profile of Shockley, John Horgan in the *San Mateo Times Weekend* cites this example of verbal ponderosity, from a letter by Shockley to *Presbyterian Life:*

> To me, it seems immoral not to view with concern, and perhaps not try to prevent, the birth of humans destined with high probability to feel that a malevolent conspiracy ruthlessly contrives their frustration. I am thinking here of those human beings forced by the improvidence of their mothers, and the obtuseness of society, to emerge into the world endowed with emotions, aspirations and capacity to remember, but so disadvantaged by an unfair shake from a badly loaded parental genetic dice cup that they have mental capacities frustratingly inadequate for our complex modern society.

Reflects Horgan, "A Hemingway he isn't."

Aware of the importance of making himself clear and avoiding misunderstanding, Shockley rehearses his speeches, often practicing them on his wife. He also mentally collects phrases and expressions that he believes have been effective and repeats them in subsequent speeches and articles: the "badly loaded parental genetic dice cup" concept in the preceding quotation is an example. Metaphor-collecting is common among public speakers but unfortunately in Shockley's case the effect is a thick mass of metaphors: "It is my intention to use significant members of the American press as the blocks or pulleys . . . and the First Amendment as a line upon which I shall endeavor to exert a force so as to deflect the rudder of public opinion and turn the ship of civilization away from the dysgenic storm that I fear is rising over the horizon of the future."

Shockley's efficacy as a speaker is further hampered by his tendency to read hurriedly from written — or memorized — statements. He reads partly because he has difficulty extemporizing and partly, undoubtedly, to be careful. In a television debate on the David Susskind Show, during which Shockley frequently quoted segments of his doctrine from memory, at one point he retorted to Susskind, "You're not going to put words

into my mouth that are not there in my written statements."
His collage of prepared statements and charts has unfortunate
effects on his audience. The reaction of writer Larry King,
hearing a debate at Princeton University, was that Shockley
"made such an inept presentation that he probably could not
have instructed us how to catch a bus. He asked for a bewildering number of lantern slides to be flashed across the screen —
weird, statistical, silent rantings with zigzag curves sometimes
resembling grasshoppers with their backs broken or a row of
vacuum cleaners — saying they proved this and that and what-
not. People yawned and scratched and removed themselves in
small clots and knots and bunches."

Further decreasing his efficacy, Shockley maintains a kind
of love-hate relationship with the press. On the one hand, he
wants publicity. He collects and makes hundreds of multi-
lithograph copies of news articles about himself — pro and con
— and has them filed methodically by index number in large
folders around his office. A physicist who had invited Shockley
to speak at Sacramento State College, California, recalls that
after the speech Shockley dashed out to get a copy of the *Sacramento Bee* to see what it said about him. He also got word that
a reporter had been trying to reach him. Worried that the
reporter would leave work before he reached him, Shockley
wanted to interrupt his schedule and call him immediately. His
host was hard put to convince Shockley to go back to his motel
first, as planned, and call the reporter from there.

At the same time Shockley is unusually distrustful of reporters. Intensely worried about wasting time, he puts would-be
interviewers through what freelance writer Michael Rogers
calls a "subtle and continual screening process." Preparing a profile of Shockley for *Esquire*, Rogers was first required to spend
several hours in Shockley's outer office performing statistical
exercises under the supervision of a graduate assistant. Then,
having demonstrated sufficient ability and good faith, he won an
audience with Mrs. Shockley, who talked willingly and openly,
gave him two pounds of reprints as homework, and reported a
favorable impression to Dr. Shockley. Her recommendation
prompted a call from Shockley himself, and an invitation to

his home for more exercises. There, Rogers listened to a number of tape recordings Shockley had made of earlier encounters with unsympathetic interviewers, and Shockley noted Rogers's reaction. In conversation, Shockley continued to test Rogers's interviewing performance; Rogers says, "Shockley does not hesitate to tell me when a question is 'stupid.' Later, in fact, in a late-night telephone interrogation that I fail miserably, 'stupid' will become my 'pet name.' " The test that Rogers ultimately failed went as follows:

In a belligerent forty-five minute telephone interrogation (which Shockley opens by announcing, "This is William Shockley. I am a press correspondent for the *Manchester Union Leader*"), we discuss a television program he has asked me to view: our respective responsibilities for the future of mankind, the goodness of the fundamental search for truth, the fundamental quantitative realities, and a variety of other subjects, and at length Shockley concludes, "I don't think you're playing this game tough and for keeps."
I tell him that I'm not aware it is a game.
"You're sort of flunking this one so far," he says, and then later, reluctantly, decides, "You might have been in on some interesting conference calls but I'm afraid this doesn't quite live up to that."

Shockley sometimes asks journalists to do some writing for him, implying this may be the price of a good interview. In an early interview for this book, Shockley made such a proposal, and failure to cooperate brought abrupt dismissal from his home. "You're not using me," Shockley called from the doorway; "I'm using you."

Shockley seems at best inept, at worst insulting, with reporters. According to Larry King, Shockley invited the press to the Nassau Inn before his scheduled debate at Princeton, billing the event as a "luncheon press conference." "What was wrong with that," King reports, "was that only Dr. Shockley got anything to eat. 'I don't have an expense account,' he explained to newsmen who munched candy bars or potato chips."

Reporters' difficulties with Shockley are aggravated by the fact that, wanting to present both sides of the issue, they must turn to critics often as difficult to understand as Shockley. Refutations of Shockley's position tend, from the viewpoint of the reporter, to be emotional, off-the-cuff, ill-prepared, or abstruse. At a Stanford University debate in January 1973, one of Shockley's opponents, a Stanford psychologist, countered Shockley's position, as follows:

You'll notice that in his writings he speaks to the notion of his null hypothesis being 80% geneticity. . . . He has inappropriately conceptualized, defined, and really described what in fact is happening with the null hypothesis as used in psychology. In psychology the null hypothesis is a statistical hypothesis which *a priori* states that there is no difference between or within a given population or that the difference is due to chance. Thus, looking at a normal distribution with a total probability of 1, or 100, we would then be looking, be speaking in terms of a null hypothesis with respect to 50% geneticity, and we would then be speaking to an alternative hypothesis of 80% geneticity or whatever else, or whatever number that he would see as appropriate. So you see it is also very hard for me, as a social psychologist, to deal with the entire logic of a program in which the fundamental assumptions are inappropriately drawn and erroneously applied.

A Hemingway he isn't, either.

The credibility of critics is also difficult for the press to establish, because Shockley makes a strong case that objections to his work are frequently based on prejudice and fear, not on a reasoned investigation of the research. And Shockley, who appears cool and calm in public, makes emotional critics look foolish. History has shown that the minority view in science has sometimes been proven right in the end, as minority figures like Shockley are quick to point out. In the meantime, the virulence of the attacks on Shockley, and the reluctance of professionals to do research in the touchy race area, have lent credibility to Shockley's claim that he is indeed the victim of political forces.

The influence of politics on genetics is nothing new, according to Cornell historian William Provine. Tracing the history of "geneticists and the biology of race crossing" from the nineteenth century through Jensen, Provine argues that the political beliefs of geneticists have often influenced their interpretations of the scarce data available and profoundly affected their conclusions. With little change in the data, Provine shows, between 1930 and 1950 the genetic community reversed its stand on the harmful effects of race mixing. He urges in conclusion, "I am not condemning geneticists because social and political factors have influenced their scientific conclusions about race crossing and race differences. It is necessary and natural that changing social attitudes will influence areas of biology where little is known and the conclusions are possibly socially explosive. The real danger is not that biology changes with society, but that the public expects biology to provide the objective truth apart from social influences. Geneticists and the public should realize that the science of genetics is often closely intertwined with social attitudes and political considerations."

Shockley has countered the odds against him in the press by becoming a master at obtaining publicity. He periodically manufactures pseudoevents, usually by introducing into his set of public views a new startling idea, and timing the release of the idea with care. In an interview with the *National Enquirer* in March 1973, Shockley discussed a eugenics idea he had earlier publicized, a voluntary sterilization bonus plan, only this time he added a new suggestion. The plan as he had discussed it before, and which he had been careful to call only a "thinking exercise," would give cash incentives for sterilization to poor, low-IQ citizens — for those who do not pay income tax, $1,000 for each point their IQ fell below 100. Ten percent of the bonus, he had proposed, might go in spot cash to the person who located a volunteer for sterilization, stimulating "our native American genius for entrepreneurship." To the *Enquirer* reporter, Shockley added that the recruiters earning the ten percent might be thought of as "bounty hunters." "I conducted an evaluation by telephone with some editor friends," Shockley says, "with the conclusion that it would be effective to introduce the 'bounty hunter' phrase." When it did not appear in the final *Enquirer* story, Shockley introduced it in a press release for a June 1973 speech, and the startling "bounty hunter" idea then made local headlines.

He found another opportunity to manufacture a pseudoevent when Roy Wilkins, executive director of the National Association for the Advancement of Colored People (NAACP), commented on Shockley's theories in a speech in Indianapolis, Indiana, in July 1973. Shockley took the occasion to write to Wilkins, and to issue a press release, calling on Wilkins and other black intellectuals to help him research the question of Negro inferiority. Black leaders could, Shockley suggested, take blood samples from one or two hundred black intellectuals, to see whether, as Shockley has predicted, IQ increases one percent for every one percent Caucasian ancestry. His bizarre suggestion was heavily covered in Bay Area newspapers.

Shockley often keeps the ball rolling by writing a letter to the editor, objecting to inaccuracies and misrepresentations in an article. If the reporter replies, Shockley replies, until one simple news article can result in several days' publicity. A speech to San Francisco's Commonwealth Club, for example, covered by science editor David Perlman in the *San Francisco Chronicle*, generated a letter from Shockley, a reply by Perlman, and another letter and reply. One disgusted science editor, recognizing that Shockley often tied up space with letters-to-the-editor hassles, says he has made it a policy not to print articles about Shockley.

In spite of his hostility toward the press, Shockley has cultivated a number of sympathetic newsmen in a kind of advise-and-consent role. His wife explains that Shockley has gotten acquainted with certain newsmen, sometimes through meeting them socially, sometimes when they interviewed him, and now finds he can call on them for advice on how to handle press stories. He sometimes calls to check whether he has made a news release clear, or how best to time an announcement. The newsmen oblige him — in some cases, it would seem, out of courtesy, and in others because they believe in his cause. His relationship does not seem to be the typical favor-exchanging one, in that the reporters rarely call on Shockley for help with other stories. David Tennant Bryan, publisher of the *Richmond* (Va.) *Times-Dispatch* and *Richmond News Leader*, and William Loeb, publisher of the *Manchester* (N.H.) *Union Leader* are among his contacts, according to Mrs. Shockley.

An incident in February 1973 epitomizes Shockley's elaborate handling of the press. Shockley flew to London, where he was giving a speech for the Institute of Electrical Engineers commemorating the twenty-fifth anniversary of the invention of the transistor. While in London, Shockley was invited for cocktails by Lord Boyle, vice chancellor of Leeds University, where Shockley was scheduled to receive an honorary degree the following May. When Boyle steered the discussion onto Shockley's genetic theories, Shockley unobtrusively reached under the table and turned on the tape recorder in his briefcase. Describing the move in a later interview, Shockley called it sneaky: Boyle did not know the recorder was on until Shockley had to turn the tape over, forty-five minutes later. In the meantime, sensing a "change of tone," Shockley had asked Boyle if he would prefer to forget about the degree. Explaining that the degrees committee had not been aware of Shockley's genetics papers at the time it was decided to grant a degree, Boyle admitted, on tape, that the committee would indeed like him to forget it.

One might expect Shockley to want to keep such an unflattering story out of the newspapers, and Leeds University would have been only too happy to do so. But Shockley's reaction was quite the opposite. Excited about the possibility of a news story, Shockley made a transatlantic phone call to one of the newsmen he consults. From the tape of that phone call, which Shockley played in an interview, the conversation was long, convoluted, and confused. At one point Shockley mentioned an interview he was going to have on the David Frost Show and the newsman cautioned Shockley, reminding him that Frost was engaged to marry black singer Diahann Carroll. Shockley sounded embarrassed and hastened to say he was not a racist. It finally became apparent to the newsman that Shockley had called to ask how to get the most mileage out of the Leeds incident. Shockley proposed that he mention the situation in an interview with the *Times* (London) scheduled for the following morning. Reacting to a hint from Shockley, the newsman then agreed to ask an AP friend to cover the interview. The story thus made major headlines in London and the United States: "Transistor Man Gets Message," "Leeds Snub for Nobel Scientist."

Shockley felt Lord Boyle had played right into his hands. The unpleasant Leeds situation simply afforded him another opportunity to get his human quality views into print. He found an apt phrase to describe his philosophy about adverse publicity while rehearsing his London transistor speech at Bell Labs before he went to London. After the speech one of the workers offered him a motto. Shockley took to it immediately and quoted it frequently to newsmen during the Leeds episode: "When life gives you lemons," it said, "make lemonade."

Shockley's unusual press activity stems from an apparently deep and genuine concern for the quality of human life. His underlying worry is that the intelligence of the human population is declining because parents with low IQ are having disproportionately more children than those with higher IQ. This dysgenic effect, he believes, is most threatening in American Negroes, and could result in their "genetic enslavement," as well as an overall lowering of IQ in the American population. His alarm has led him to devote almost all of his waking time to promulgating and defending his position. (He is now professor emeritus of electrical engineering at Stanford University, with no teaching duties.) He is convinced he is intellectually and morally justified in his devotion to dysgenic concerns:

I believe my actions in raising these questions are like those of a visitor to a sick friend who urges a thorough diagnosis, painful though the diagnosis may be, so that remedial steps may be based on objectively established facts and sound methodology. To fail to raise these unpopular questions because of fear of the resentment towards me that may ensue is an irresponsibility I am not willing to have on my conscience. I believe and hope that my determination to see that these questions are faced and answered may be the greatest contribution anyone can make to American Negro welfare for the next generation.

Keenly aware that his time is limited, he declares his ultimate goal in a "death postulate": "During the last rational five minutes of my life, should I happen to have my intellectual powers intact, I hope to consider that by demanding objective inquiry and open discussion of human quality problems I have used my capacities in keeping with the objective, like that of Nobel's will, of conferring greatest benefit on humanity." Shockley's religious position, his wife says, fluctuates between atheism and agnosticism.

Mrs. Shockley, who became the scientist's second wife in 1955, provides strong moral support. She accompanies him on his travels and serves as a kind of executive secretary in his office. Before marrying Shockley, the former Emmy Lanning was a psychiatric nurse, with a master's degree from Columbia University nursing school. The couple met by accident. The then Miss Lanning was invited for dinner at a friend's house in Washington, D. C., one night, Shockley the next, but Shockley arrived a night early. The hostess invited him to stay anyway and they all had dinner together. At first she did not like Dr. Shockley, Mrs. Shockley recalls, because he ignored her totally, but at dinner she piped up to disagree with him about the study he was describing, and he took notice. She has become passionately involved in his dysgenics crusade, although her devotion is based not so much on her own convictions as on loyalty to his: "Dr. Shockley can get involved in something called mankind; I can't. . . . It overwhelms me to think about mankind. . . . I'm trying to help him do what he wants to do, and my relationship with him is important, and when it gets not to be important any longer, then I'll have to do something else about this." She donates nearly all her energies to the crusade, working long, exhausting hours as an unpaid office assistant, with duties ranging from maintaining an elaborate filing system to running errands, screening calls, and serving as a guinea pig when her husband wants to see if his lessons are clear. Shockley also hires student assistants (who, he stipulates, must have Scholastic Aptitude Test scores over 700). The Shockleys have an attractive homey ranch house in the Stanford faculty section, but do very little socializing. Reflecting on accusations that Shockley is a Fascist, a racist, a monster, Mrs. Shockley says, "Someday we may actually be terribly alone."

Shockley comes to his obsession with the dysgenic question

late in life and as the result of a number of experiences. As his "death postulate" implies, he was influenced by his receipt in 1956 of the Nobel Prize, an honor that so often confers on its recipients a sense of social responsibility and mission. It also confers publicity, which, according to a friend who worked with Shockley at Bell Laboratories, Shockley has always enjoyed. In the flush of prize publicity, Shockley took the opportunity to give a number of speeches, more than his co-discoverers, and these were covered by the press. The pattern of seeking publicity that he uses today is nothing new, according to one co-worker; he is a person, the colleague explained, "who thinks you make progress by using publicity."

It was also not new for Shockley to venture outside the realm of physics. In World War II, he conducted operations research for the navy, an experience he considers a key contribution to his later genetics study. His mimeographed biographical handout says:

My qualifications to reach conclusions in the field of human genetics are not those of a geneticist, a psychologist or an anthropologist, nor have any of my statements suggested that I thought I was so qualified. I do, however, bring the qualifications of a scientist, an educator, an engineer, and specifically, my operations research experience of World War II. The phrase 'operations research' was invented in World War II to describe scientists working with military commanders to analyze statistical and scientific aspects of combat operations. My activities were concerned with antisubmarine warfare and radar bombing techniques and my contributions in these unfamiliar fields brought me the Medal for Merit, the highest civilian decoration. I regard my role in respect to human genetics as being professionally similar to my wartime experiences in the sense that detailed knowledge of the intricacies of the field may even distract attention from the central issues to important but irrelevant details. It is my experience in operations research that I believe best qualifies me to reach the conclusions and recommendations that I ... state today.

In a 1954 study of scientific productivity, conducted by Shockley for the Department of Defense, similar skills were used. Shockley observed in the study that there are exponential differences in rates of publication among scientists, and he argued that military and other government laboratories should pay higher salaries to attract high producers. These exceptional scientists, Shockley speculated, have higher "brain capacity," great ability to handle the relationships among ideas.

With the 1954 Pentagon study, conducted while on leave from Bell Laboratories, outside activities began to predominate in Shockley's life. The next year he quit his job at Bell, where he had worked since he received his Ph.D. from Massachusetts Institute of Technology in 1936, and where he had done his Nobel Prize–winning work on the transistor, and went into business for himself. He began at Raytheon Company, where he proposed to establish a semiconductor firm. His terms, however, were that Raytheon guarantee him $1 million over a three-year period. This was "hardly unreasonable by today's standards," *Electronic News* points out, "but the Raytheon management sixteen years ago couldn't see it." Shockley therefore left Raytheon after only one month. His next step, with the backing of Beckman Instruments, was to establish the Shockley Semiconductor Laboratory in Palo Alto, California, to develop and produce the new transistor and related devices.

But Shockley as a businessman had personnel problems. Coworkers at Bell Labs say he had previously had difficulties as a department head at Bell: many of his people had quit because they couldn't get along with him, and others he had fired. At his own company, the personnel problem was fatal. Eight key employees left together in 1957, forming Fairchild Semiconductor. "Shockley Transistor never recovered from the blow," according to *Electronic News*, "although the company staggered along, through three owners, until mid-1968. Beckman gave up and sold the company to Clevite, which gave up and sold it to ITT. ITT gave up, couldn't sell it, and shut it down."

Meanwhile, Shockley had taken an appointment as professor of engineering science at nearby Stanford University in 1963. There his teaching experiences led to another kind of "outside" activity: "thinking about thinking." Based on his personal experiences and observations of how creative breakthroughs are made, he began theorizing about scientific thinking and how it could be improved. He developed elaborate "science-thinking tools" and concepts such as the Creative Search Pattern, Scientific Logical Structure, and Qualified Law Form. In the mid-1960s, concurrent with his early dysgenics campaigning, he applied some of his ideas in six Bay Area junior high and high schools as part of a study supported by a grant from the U. S. Department of Health, Education and Welfare. His college textbook, *Mechanics*, coauthored by Walter A. Gong, is based on the same approach.

Shockley's physics research in this later period was sound but of course sparse. According to a fellow Stanford University engineering professor, Shockley made an "important conceptual contribution" by recognizing the potential of new magnetic "bubble devices" for computer memory technology and persuading Bell Laboratories to look into the idea. Prompted by rumors that he was getting senile, he also pursued a problem of "hidden momentum" forces in electromagnetic theory, clearing up a point that had been glossed over since Einstein. Using a different approach, the same discovery was made simultaneously and independently as part of a large project at the Massachusetts Institute of Technology by Paul Penfield, Jr., and Herman A. Haus, described in their book, *Electrodynamics of Moving Media*. Shockley's work was sound and original, according to physicists working in the field, although "not earth-shaking," and "out on the edge of things." Shockley hit upon the idea during a course he was teaching at Bell Labs in which he was applying his science-thinking ideas, and he and a graduate student published the resulting technical papers from this science-thinking point of view. Colleagues ignored the "kooky" point of view as unnecessary and unimaginative, but they feel he is a "first-class physicist." His image in the physics community, they concede, has been tainted by the genetics controversy, but it had been earlier anyway when Shockley formed his semiconductor company and went on the road around the country, "hustling his products."

Shockley's best-known "outside" activity, his campaign on human genetic quality, began to demand most of his time in the later 1960s. The process had been gradual. Shockley was by temperament likely to take an interest in human quality. Born in 1910 while his American parents lived in London, he was raised in California, received his B.S. from Cal Tech, then a Ph.D. in physics from M.I.T. Friends considered him a "genius," and highly competitive. Describing an experience when he was a student at Hollywood High School, Shockley told a biographer, Shirley Thomas, "I recall on the occasion of an examination in mathematics, I was not sure that I had done as well as I might have. I was especially concerned about my relative grade to that of another student — one who was not at all remarkable. All weekend I worried over whether I had done better than he. This seems so trivial now, but it was important to me then." Shockley also told Thomas of his concern that a "cult of mediocrity," a prejudice that no one can really be much more able than another, hampered this country from fulfilling its potential. A similiar interest in fostering people with higher mental potential is evident in his 1954 Pentagon study of scientific productivity. Shockley is a registered Republican, and has been "strongly right wing" since Bell Labs days, according to a longtime friend.

In the early 1960s, certain incidents reached Shockley's attention that, when combined with his previous outlook, produced his dysgenic worry. The first, according to Shockley, was in 1963; he was asked, as a Nobel laureate, to sign an endorsement supporting the position of the Society for Humane Abortion. "It took me a year to reach the view that a woman should have a right to an abortion much as she does for a surgical operation." Also in 1963, Shockley was impressed by a news story about a San Francisco delicatessen proprietor who was blinded by a hired acid-thrower. "He [the acid-thrower] was one of seventeen children, almost all illegitimate," Shockley recalls. "His mother could remember the names of only nine of her children. She had an IQ of 55. The teen-ager, Rudy Hoskins, nicknamed 'the brute,' had an IQ of 65." Two years later, describing his reaction to the story, Shockley said, "If that woman can produce seventeen children in our society, none of whom will be eliminated by survival of the fittest, she and others like her will be multiplying at an enormously faster rate than more intelligent people do. Is she an isolated statistic? Who knows? For myself, I fear it is not an isolated statistic." Shockley also mentions frequently a 1966 account in *U. S. News & World Report* of what he calls "Denmark's sterilization programs with their eugenic implications." According to the *U. S. News* article, Denmark has instituted a system of free sterilizations, with inducements containing varying amounts of coercion, for criminals, psychotics, alcoholics, the chronically jobless, the severely retarded, and couples with a family history of hereditary disease. The article is enthusiastic, and Shockley speculates, "The rising per capita homicide rate of Washington, D. C., is fifty times Denmark's falling one. Dysgenics?"

Shockley's early speeches and interviews on the subject were centered on the overall population, without regard to race. The situation changed, however, within a year of the time he began speaking on eugenics and dysgenics. News accounts of a speech at Gustavus Adolphus College in St. Peter, Minnesota, brought him to the attention of *U. S. News & World Report* magazine. Having been misquoted at Gustavus Adolphus, Shockley consented to a *U. S. News* interview but required a written agreement, stipulating that the article not be released until he had approved it in final, edited form. When the interviewer asked Shockley a question about Negroes, Shockley answered cautiously but candidly:

Q: To what extent may heredity be responsible for the high incidence of Negroes on crime and relief rolls?

A: This is a difficult question to answer. Crime seems to be mildly hereditary, but there is a strong environmental factor. Economic incompetence and lack of motivation are due to complex causes. We lack proper scientific investigations, possibly because nobody wants to raise the question for fear of being called a racist. I know of one man who is writing a book in this area, and I'm not sure he'll finish it because the subject is so touchy.

But let me say what I find in my own reading:

If you take the distribution of IQ's of Negroes, and compare it with that of whites, you are going to find plenty of Negroes who are superior to plenty of whites.

But, if you look at the median Negro IQ, it almost always turns out not to be as good as that of the median white IQ. At least, this is so in the U.S. How much of this is genetic in origin? How much is environmental? And which precise environmental factors are to blame? Again, a "controlled" program of adoptions might give answers.

Actually, what I worry about with whites and Negroes alike is this: Is there an imbalance in the reproduction of inferior and superior strains? Does the reproduction tend to be most heavy among those we would least like to employ — the ones who would do least well in school? There are eminent Negroes whom we are proud of in every way, but are they the ones who come from and have large families? What is happening to the total numbers? This we do not know.

Shockley's first comparison of races in public — his remarks indicate he had considered the matter privately for some time — had far-reaching repercussions, including what he calls "an experience that led me deeply into racial matters." In the process of approving the *U. S. News* interview, Shockley sent a draft version of it to "several dozen people in relevant scientific disciplines." One copy "accidentally fell into the hands" of the editor of *Stanford M. D.*, an alumni magazine of the Stanford University medical school. The editor made "several helpful editorial suggestions that led me to release it," and obtained Shockley's permission to reprint the interview. When the article appeared in *Stanford M. D.*, the Stanford Medical School genetics faculty sent a joint letter to the magazine, condemning Shockley's views: "This kind of pseudo-scientific justification for class and race prejudice is so hackneyed that we would not ordinarily have cared to react to it. . . . We deplore his innuendos. . . . It falls between mischief and malice to make such a prejudgment in these terms."

Criticism only spurred Shockley on, because it suggested that eugenics and race were "taboo" subjects. According to Shockley, he was frequently urged in 1966 to continue his human quality work but to leave out the touchy racial aspects. He says that he wrote Harvey Brooks, Chairman of the National Academy of Sciences (NAS) Committee on Science and Public Policy, for advice in responding to the geneticists' letter, and he quotes Brooks as replying, "Ignore the question of race; otherwise you are simply guaranteeing yourself against an objective audience." Resenting such restrictions in science, in 1966 Shockley made a decision: "By campaigning against these taboos as a Nobel laureate, I would serve the objective of the 'greatest benefit on mankind' phrase in Nobel's will. I also felt an obligation: I would risk less than would others sharing my convictions but not my academic security." Adding to his frustration — and zeal — about "taboos," Shockley has occasionally learned over the years of professionals who support him privately, even passionately, but not publicly. From the letter of an M.I.T. professor, a member of the National Academy of Sciences, who had visited him in 1972: "Sir, for the first time in my life I felt to be in the presence of a true prophet: Jeremiah, Christ, Galileo, Shockley. . . . I have no doubt that you will go on in history not only for the solid state, but for this matter as well." The letter compared the NAS's disapproval of Shockley's position to the church's attack on Galileo.

Concentrating on the race area, where criticism centered, Shockley searched out data to support his original points. Backed primarily by private donors, he began devoting full time to analyzing IQ research. At first, he spent $5,000 a year out of his own pocket for expenses, but as publicity grew, checks from sympathetic readers began to come in.

His interpretations of the data strengthened and magnified his views. In contrast to his caution in the 1965 interview, in 1970 he told the National Academy of Sciences: "My research leads me inescapably to the opinion that the principal cause of our American Negro problems is racially genetic. I call it an opinion rather than a proof largely because it simply does not receive objective examination by competent scientists with appropriate qualifications. . . . Diagnosis will, I believe, confirm that our nobly intended welfare programs are promoting dysgenics — retrogressive evolution through the disproportionate reproduction of the genetically disadvantaged." His work also took him further out on a limb. By 1972 he asserted, for instance: "Perhaps nature has color-coded groups of individuals so that we can pragmatically make statistically reliable and profitable predictions of their adaptability to intellectually rewarding and effective lives."

Shockley is the first to point out that he welcomes adversity and controversy — because he recognizes its news value. *San Mateo Times* writer John Horgan experienced this example:

Just before he is to speak to a student gathering at UCLA, he mentions that it might be a good idea to have the Associated Press cover the meeting since, in his view, there is a chance that it might be disrupted. With some laughter, he adds that, "If the thing does get busted up, it'll mean that much more coverage — but I'm not planning it that way." Shockley is correct: The encounter does engender some noise on the part of some students and the AP is indeed there to record it, albeit none too precisely, or completely.

Shockley has been hanged in effigy; his classroom has been invaded by students in Ku Klux Klan robes; "Sterilize Shockley" is painted on campus walls; but Shockley is not deterred. He uses each incident that finds its way into the news to reiterate and explain his views.

Unmoved and uninfluenced, Shockley also uses attacks to bolster his belief that he is a victim of "Lysenkoism." "Most of the attacks," he told an interviewer in 1972, "are not particularly disturbing because they fall so neatly into a pattern of interpretation I've already set up."

Shockley is even able to "make lemonade" out of *not* being heard. Several appearances have been canceled because opponents threatened to disrupt a scheduled speech, or actually broke up the lecture while in progress. Two incidents in November 1973, a canceled debate with Roy Wilkins at Harvard University, and a disrupted speech at New York's Staten Island Community College, caused *Time* to reflect: "The irony of the Shockley case is that a questionable, perhaps even pernicious, doctrine is probably receiving more publicity by not being heard than open debate would give it."

The Visibility System

Having only a few visible scientists, if that's the way it works out, is not good, but it's better than having no visible scientists.

— CARL SAGAN

The first time Philip Abelson was invited to appear before a congressional committee, he prepared his presentation carefully. When he arrived, he found that Edward Teller was among the witnesses. "Indeed, he was the lead-off speaker," Abelson recalls. "When his testimony was completed, congressmen swarmed around, having their pictures taken with him. The press clustered about, firing questions. When Professor Teller left, almost all the congressmen and the press vanished. The testimony of subsequent witnesses had as much substance as that of Teller, but their voices echoed in an empty room."

The incident is incredible to most scientists since, although Teller is indeed a celebrity, remembered as the "father of the H-bomb," Philip Abelson is an influential figure in his own right, editor of *Science* magazine and president of the prestigious Carnegie Institution, Washington, D. C. The situation is rather like one celestial body eclipsing another.

Many a scientist has had an experience similar to Abelson's. Echoing the same phenomenon, *Boston Globe* science reporter Robert Cooke observed the effect of Carl Sagan's charisma at a recent AAAS meeting: "Sagan, usually wearing a turtleneck shirt that seems to be almost a trademark, drew standing-room-only crowds when he spoke. After he finished, many, especially the young, left before the program ended. The other scientists usually ended up speaking mostly to their fellow scientists."

It is a different system, the visibility system, and one that bothers not just spurned scientists but social critics and conscientious journalists as well. Do the visible scientists have too much influence on the mass media and public opinion? The problem has two parts. First, visible scientists may command too much attention from the press, because they are more accessible and more quotable than other scientists. Second, they may command too much authority, because they speak as scientists yet their topics are outside their area of technical expertise. The result could be that visible scientists get a disproportionate amount of coverage from the media, and have an overrated credibility with the public.

Visible scientists have been subjected to some stinging criticism for pushing the limits of their credibility. Arthur Herzog brands visible scientists, and others like them, "Anything Authorities":

The Anything Authority is someone whose credentials in one field are taken as valid for others — often many others but almost always including politics. Examples are Dr. George Wald, the biologist; Dr. Benjamin Spock, the pediatrician; Jane Fonda, the actress; and Dr. Linus Pauling (who has said of President Nixon: "For years I have studied insanity. I saw the eyes on television, and *there* is madness, paranoia"), the chemist turned psychologist.

The trouble with an Anything Authority is not that he takes a position or works for a cause but that he seldom seems to apply the same standards of research and documentation to the field in which he is not an expert that he would to his own. . . . The opinions of Dr. Pauling, twice winner of the Nobel Prize, must always be respected, and yet it is difficult to imagine Pauling discussing the double helix, which he helped discover, in quite the glib manner he used in talking about Nixon's paranoia. For in his own discipline the authority must prove his contention — a requirement that vanishes when the topic is Anything.

When an Anything Authority becomes successful, he joins the Permanent Rotating Panel Show on television. Besides fame as an Anything Authority, there are two criteria for membership on the PRPS. One is that the guest must get a reaction from the audience

— whether cheers or boos doesn't matter — and the other is that he must *never* be stuck for an answer.

Being an Anything Authority is big business these days. They are quoted in magazines, called upon to write book reviews, invited to join boards of directors to add panache, given sizable per diems as consultants, and used for all manner of fund drives to benefit any cause from politics to private schools. The whole phenomenon results from the hope that somebody still knows something outside the narrow fields of specialization — despite the mounting evidence to the contrary.

Compounding the problem is the danger that visible scientists are viewed not only as individual experts but also as representatives of the scientific community as a whole. Their credibility will be even more exaggerated if the public assumes their colleagues stand behind them.

Scientists in general have had high credibility, benefitting from the traditional image of scientists as disinterested, specialized professionals. Both scientists and laymen have tended to assume that because scientific research is always verifiable, scientific researchers are always objective. In spite of ample evidence to the contrary, the myth of scientists' objectivity persists, and the real and important role of creativity, subjectivity, and serendipity in the discovery process is downplayed. Junior high school students are taught the "scientific method," although successful scientists often protest that science simply does not work that way; says Albert Szent-Gyorgyi, a Nobel laureate who has made major contributions in several fields: "Naturally, like all other scientists, when I publish or speak about my work, I like to make it appear as though it has been one straight line, one preconceived logical unit. But while I work I usually do not know where I am going. I just follow hunches."

As long as scientists are credited with special objectivity, they have a special credibility problem when they discuss public issues. The solution, some have suggested, is for scientists to make public statements only on matters within their scientific specialty. Another suggestion has been for scientists, when they make a statement, to state clearly whether they are speaking as "scientists," on a subject in which they have specialized technical information, or as "concerned citizens," with the same strong opinions and biases as any other member of the involved public. A third proposal has been a court-like adversary procedure, in which scientist-judges would determine the technical facts of an issue after hearing from scientist-advocates representing all sides.

For the highly visible scientists, the solution would seem to be for them to continue to be very much themselves. Flamboyant, passionate, hyperbolic, political, they defy audiences to think of them as coolly objective, unbiased, technical. Thus they forfeit a claim to scientific objectivity and the credibility that goes with it, using scientific facts but not scientific image. They are welcomed by the public not as scientists but as authors, television personalities, mavericks, heroes, and antiheroes of a technological and electronic age.

There is no question that visible scientists are making public statements based on opinions outside their specialties of lepidopterology, biochemistry, or plant physiology. Paul Ehrlich says so explicitly, in an "Author's Note" at the beginning of his second paperback, *How to Be a Survivor:*

> Scientists in our society are often criticized for being too narrow, but when they step outside the narrow field of their training they are invariably told they should stick to their specialty. We prefer to risk the latter criticism. In many areas of this book, we have gone beyond the boundaries of our formal training to try to seek solutions to human problems. We see no other course than for scientists in all fields to do the same — even at the risk of being wrong.

Ehrlich, who has given the situation a good deal of thought, sees several dangers. "Some scientists," he has said, "because of the field they are in, or some honor that they receive, or because of a personal bent for communication, find themselves in a position to influence social policies directly, rather than indirectly through their work. They become public figures, make public pronouncements, and are asked for advice by government officials. This presents some massive problems for them, both as scientists and as individuals."

The visible scientist faces the dilemma, Ehrlich continues, of

"how much generalization and drawing of conclusions is legitimate (in contrast to 'letting the facts speak for themselves') in
order to get the message across?" Ehrlich believes he made a
number of errors of judgment in the first edition of *The Population Bomb*. "For example, at the time I thought that American leadership and even diplomatic pressure would help change
the policies towards population control in underdeveloped countries. I now think that the United States should play a much less
active role." This change in viewpoint is reflected in the revised
edition of *The Population Bomb*.

Most difficult, Ehrlich feels, is the problem of how to present
relatively complex scientific matters to the lay public, how to
make them comprehensible without distorting them. A scientist's projections and predictions seem to be particularly subject
to misunderstanding. At the time of the Amchitka underground
nuclear test, for example, many scientists, including Ehrlich,
opposed the test, for a number of complex reasons: the probability of earthquakes occurring from the blast was small, but if
they should occur, with a possible release of radio-activity and
disturbance of the area's ecosystems, the result would be disaster — the "downside risk" was enormous. Scientists were against
the test, in other words, because, although the possibility of a
disaster was slight, the consequences if it should occur were
great. After the test, Ehrlich laments, the press reported that the
predicted disasters did not occur, and that the Atomic Energy
Commission's action was vindicated. "In my view," says Ehrlich,
"that was equivalent to praising the wisdom of someone who has
just played a round of Russian roulette and failed to blow his
head off."

Ehrlich's conclusion, however, is that a scientist must proceed with generalization in spite of the risks. "I've said lots of
things in the course of this whole business that I no longer think
were completely right or that I wish I'd done something different; I think one of the things you learn in population biology
and in science in general, if you watch carefully, is that if you
have to be 100% right in your mind, absolutely certain, before
you do anything, you are absolutely paralyzed and don't take
any action. So I think you've got to realize there's going to be

a certain amount of error." But he makes his errors with a consciousness of his role as a public figure and a sense of responsibility — "the more successful you are, the more responsibility you have."

Of greater concern than visible scientists' personal credibility is their influence via the media. Charismatic and articulate, the visible scientists bias the news: celebrities always do. The press is addicted to the visible scientists, and vice versa. The comfortable symbiosis is a source of uneasiness even to science reporters and editors. In an editorial in *Science News*, Kendrick Frazier writes: "There is legitimate concern whether the visible scientists, with their special skills, and often for quite respectable causes, unduly influence the amount and balance of coverage in favor of their positions on issues of science, technology and society. There are strong arguments either way. . . . In any case, the public interest is served when gatekeepers of the media ask themselves whether the views of a 'visible scientist' are featured because the visible scientist is more accessible and more articulate than other scientists with perhaps different views."

Media prowess gives visible scientists a number of advantages, including leverage in Congress. According to Howard J. Lewis, director of the Office of Information of the National Academy of Sciences, visible scientists play a key role in congressional hearings. Visible scientists are invited to give testimony at hearings, because they make good copy; they bring publicity to the hearings, which is exactly what the congressman wants. The scientist is, then, chosen for his visibility, and in turn the hearings increase his visibility.

As the press seeks to provide "balanced" coverage on issues, it needs counterweights to offset celebrity charisma. In the case of visible scientists, however, there are some counterbalances. As outsiders, visible scientists are often battling against strong cultural traditions, entrenched government policies, major industrial interests. They are Davids to institutional Goliaths. And one of the Goliaths is the scientific community itself. Enmeshed in a complex interdependency with government, the military, industry, universities (the federal research and development budget is now well over $20 billion a year), the scientific community has had a vested interest in influencing the flow of science news to the public. Emphasis has been on public relations, not public information. Socially alienated but politically dependent, the community has felt it needed an "objective" image, free of messy controversies or human biases, in order to merit special autonomy at home and special credibility in Washington.

The visible scientists reject the control of the scientific community, and offer a different view of science. An ironic extension of the scientific community's struggle for autonomy, they are in a very real sense independent. They resurrect for public view the intuitive side of scientific discovery and the subjective side of scientists. "Thank goodness we have them," says *New York Times* science editor Walter Sullivan, "for otherwise the public understanding of science would be far poorer."

"Politics," says science writer Victor Cohn, "is mainly a response to the pressing and bewildering advances of science and technology and the social changes they work." There has come to be science in every aspect of politics, and politics in every aspect of science. For the public to depend solely on a single viewpoint from the scientific community would be as perilous as depending solely on the White House press secretary for political information.

The public needs alternatives, and the visible scientists are providing one. They are taking a tentative step toward a more realistic image of science, a recognition of science's strengths and weaknesses — better in the long run for science as well as for society.

Notes

Quotations not otherwise credited in the text or in these chapter notes are taken from the author's interviews and correspondence.

INTRODUCTION

page 3, popular image of the scientist: See George Basalla, "Pop Science: The Depiction of Science in Popular Culture," in G. Holton and W. A. Blanpied, eds., *Science and Its Public* (Boston: D. Reidel, 1976); David C. Beardslee and Donald D. O'Dowd, "The College-Student Image of the Scientist," *Science* 133 (March 31, 1961), pp. 997–1001; Theodore Berland, *The Scientific Life* (New York: Cowan-McCann, 1962); Bernice T. Eiduson, *Scientists: Their Psychological World* (New York: Basic Books, 1962); G. Ray Funkhouser and Nathan Maccoby, "Communicating Science to Non-Scientists, Phase I," Institute for Communication Research, Stanford University, Stanford, Calif., 1970; Philip Hills and Michael Shallis, "Scientists and Their Images," *New Scientist* 67 (August 28, 1975), pp. 471–475; Margaret Mead and Rhoda Metraux, "Image of the Scientist among High-School Students," *Science* 126 (August 30, 1957), pp. 384–390; Mitchell Wilson, *Passion to Know: The World's Scientists* (Garden City, N.Y.: Doubleday, 1972); Stephen B. Withey, "Public Opinion about Science and Scientists," *Public Opinion Quarterly* 23 (Fall 1959), pp. 382–388.

page 4, definitions: The term "scientist" is defined broadly in this book, in some cases including engineers, social scientists, and science-trained administrators, as is commonly the practice in studies of public understanding of science. Such a public-oriented definition, of course, does not always coincide with the concept of a scientist held by scientists or social scientists.

page 4, past eras of visibility among scientists: Examples can be found in Ronald C. Tobey, *The American Ideology of National Science 1919–1930* (Pittsburgh: University of Pittsburgh Press, 1971); Alice K. Smith, *A Peril and a Hope: The Scientists' Movement in America 1945–47* (Cambridge, Mass.: MIT Press, 1971); Paul Gary Werskey, "British Scientists and 'Outsider' Politics 1931–1945," in B. Barnes, ed., *Sociology of Science* (Middlesex, England: Penguin, 1972).

page 5, "cosmic overwhelm" quotation: From Diane Ackerman, *The Planets: A Cosmic Pastoral* (New York: Morrow, 1976).

page 6, McLuhan quotation: From Marshall McLuhan and Quentin Fiore, *The Medium Is the Massage* (New York: Bantam, 1967), p. 1.

page 6, Bergman quotation: From Jules Bergman, Remarks delivered at the seminar "Science Reporting Via Television and How Can It Be Improved?" American Association for the Advancement of Science, Washington, D.C., December 1972.

page 7, celebrification in the media: See Edwin Diamond, "Would You Welcome, Please, Henry and Liv and Jackie and Erica!" *Columbia Journalism Review* 14 (September–October 1975), pp. 42–46.

page 7, estimate of newspaper space: From George Gerbner, "The Press and the Dialogue in Education," *Journalism Monographs* 5 (September 1967); quotation is from pp. 5–6. Similar estimates of space for science news appear in William J. Paisley, "The Flow of (Behavioral) Science Information: A Review of the Research Literature," Institute for Communication Research, Stanford University, Stanford, Calif., 1965; G. Ray Funkhouser and Nathan Maccoby, "Communicating Science to Non-Scientists, Phase I," Institute for Communication Research, Stanford University, Stanford, Calif., 1970; and James A. Larson, "Science, Communications, Society," University-Industry Research Program, University of Wisconsin, Madison, April 1972.

page 7, "front page" quotation: From Jules Bergman, Remarks delivered at the seminar "Science Reporting Via Television and How Can It Be Improved?" (cited above).

page 8, size of scientific community: Figure of 1.7 million is from *Science Indicators 1974*, Report of the National Science Board, National Science Foundation, 1975. About two-thirds of the 1.7 million are engineers; 15 percent (245,000) have doctorates.

page 8, influence of visible scientists on science news: In a recent survey, over two-thirds of the science writers who responded indicated that "popular science-critics such as Commoner, Mead, Ehrlich" had most influenced mass media in relation to science and technology; see Sharon Friedman, "Changes in Science Writing since 1965 and Their Relation to Shifting Public Attitudes toward Science," Master's thesis, School of Journalism, Pennsylvania State University, 1974.

page 8, estimate of congressional bills related to science: From Benjamin S. P. Shen, "Science Literacy," *American Scientist* 63 (May-June 1975), pp. 265–268.

page 8, Bronowski quotation: From Jacob Bronowski, *Science and Human Values* (New York: Harper & Row, 1956), p. 13.

page 8, selection of visible scientists: As the term "visible scientists" implies, scientists were selected for the book on the basis of one characteristic: their visibility to the general public. Thus the concept of visible scientists is not the same as that of influential scientists; visible scientists are influential, but not necessarily vice versa. (In fact, scientists who are influential as government advisors are rarely visible today.) Visible scientists are also not generally "public interest" scientists, working with citizens' environmental and consumer groups. Nor do the visible scientists necessarily include the scientists who are widely known within their professional field, such as Richard Feynman or Philip Morrison in physics, Arthur Kornberg or Salvador Luria in biology.

The two surveys that provided guidelines for selecting visible scientists were a survey of science-news experts in August 1972 and a survey of college students in March 1973. In the first survey, twenty-four newspaper, magazine, television, and university science journalists suggested names of widely known scientists. In the second survey, journalism students in twelve universities around the United States were asked to identify the names that had been proposed in the first survey. The twenty scientists most widely known to the students at that time were: B. F. Skinner (correctly identified by 82% of the students), Margaret Mead (81%), Jonas Salk (78%), Wernher von Braun (56%), Linus Pauling (50%), Paul Ehrlich (49%), Isaac Asimov (47%), Jane Goodall (37%), Albert Sabin (28%), William Shockley (21%), Noam Chomsky (20%), Barry Commoner (19%), James Watson (15%), Edward Teller (14%), René Dubos (14%), Glenn Seaborg (11%), Arthur Jensen (8%), Harold Urey (4%), George Wald (4%), Joshua Lederberg (4%). By a looser definition of "scientist," including doctors, inventors, and others, the following were also widely known among the students: Benjamin Spock (95%), Timothy Leary (88%), Jacques Cousteau (88%), Christiaan Barnard (79%), Joyce Brothers (77%), Buckminster Fuller (44%), Daniel Patrick Moynihan (38%), Michael De Bakey (30%), Edwin Land (15%). Results were of course influenced by the time of the survey (1973), the unusually high education level of the students, and other factors. The small proportion of women and minorities among visible scientists probably reflects the small proportion of these groups among scientists in general; see, for example, Joseph McCarthy and Dael Wolfle. "Doctorates Granted to Women and Minority Group Members," *Science* 189 (September 12, 1975), pp. 856–859.

For further information on the process of selecting visible scientists for the book, see the author's Ph.D. dissertation: Rae Goodell, "The Visible Scientists," Department of Communication, Stanford University, Stanford, Calif., 1975.

THE TRIBAL SCIENTIST

page 11, chapter opening quotation: From Marshall McLuhan and Quentin Fiore, *The Medium Is the Massage* (New York: Bantam, 1967), p. 10.

page 11, Ehrlich "The battle" quotation: From Paul R. Ehrlich, *The Population Bomb* (New York: Ballantine, 1968), p. 11.

page 12, Ehrlich "I get away" quotation: From David M. Rorvik, "Ecology's Angry Lobbyist," *Look* 34 (April 21, 1970), p. 42.

page 13, Ehrlich quotations: "Then you really think" is from "Man Is the Endangered Species," *National Wildlife* 8 (April-May 1970), pp. 38–39; "Give your child an IUD" is from Paul R. Ehrlich, *The Population Bomb* (New York: Ballantine, 1968), p. 182; "Nothing distresses the victims" is from Paul Ehrlich and Richard Harriman, *How to Be a Survivor* (New York: Ballantine, 1971), p. 137; "Saying that the population" is from Paul Ehrlich, "World Population: Is the Battle Lost?" *Reader's Digest* (February 1969), pp. 137–140; "In the case of apparently" is from "The Political and Social Responsibilities of Scientists" (paper presented at the American Physical Society meeting, January 1972); "No one today" is from "The Environmental Crisis," an unpublished speech manuscript; "There are 3.6 billion" is from "The Population Bomb," an unpublished speech manuscript; "People haven't wanted" is from Jeffrey Kirsch, interview with Paul Ehrlich, "San Diego Science Scene," KPBS television, San Diego, Calif., May 1975.

page 14, Ehrlich "You do your best" quotation: From "The Political and Social Responsibilities of Scientists" (cited above).

page 14, Ehrlich "Not-too-bright . . . mental defectives" quotation: From Jeffrey Kirsch, interview with Paul Ehrlich (cited above).

page 14, Ehrlich Bulletin article: Paul R. Ehrlich, "The Benefits of Saying YES!" *Bulletin of the Atomic Scientists* 31 (September 1975), pp. 49–51; "Acknowledgements" quotation is from p. 51. Dyson article is Freeman Dyson, "The Hidden Cost of Saying NO!" *Bulletin of the Atomic Scientists* 31 (June 1975), pp. 23–27.

page 16, Ehrlich Chemistry and Stanford Review articles: Paul R. Ehrlich, "1984," *Chemistry* 38 (August 1965), pp. 12–17; Paul R. Ehrlich, "The Biological Revolution," *Stanford Review* 67 (September-October 1965), p. 20–22.

page 19, list of visible scientists: The list that Dubos and Mead commented on is a list of thirty-nine scientists derived from a 1972 survey of science-news experts; see last note for *page 8*.

page 19, importance of relevance in science news: See Sharon M. Friedman, "Changes in Science Writing since 1965," Master's thesis, School of Journalism, Pennsylvania State University, University Park, 1974:

page 20, "world's foremost authority" quotation: From *Current Biography*, 1967, s.v. "van Lawick-Goodall, (Baroness) Jane."

page 21, Bergen article: Candice Bergen, "With Jane Goodall in Africa," *Ladies' Home Journal* 92 (February 1975), p. 32ff.

page 21, "ethereal" quotation: From Charlotte K. Beyers, "Beauty and Her Beasts," *Saturday Review of the Sciences* 1 (February 1973), p. 34.

page 22, McGrady quotation: From Patrick M. McGrady, Jr., *The Love Doctors* (New York: Macmillan, 1972), p. 274.

page 23, importance of controversy in science news: See Bruce Jon Cole, "Science Conflict: A Content Analysis of Four Major Metropolitan Newspapers, 1951, 1961, 1971," Master's thesis, School of Journalism and Mass Communications, University of Minnesota, Minneapolis, 1974.

page 23, McCabe quotations: From Mary Hager, "Professor Leaps from Butterflies to Birth Control," *Washington Post*, February 22, 1970, p. D5.

page 23, Strohm quotation: From John Strohm, "Dr. Ehrlich: Prophet or Doom Monger?" *National Wildlife* 8 (October-November 1970), p. 17. The interview he refers to is "Man Is the Endangered Species," *National Wildlife* 8 (April-May 1970), pp. 38–39.

page 24, Time *quotation:* From "Skinner's Utopia: Panacea or Path to Hell?" *Time* 98 (September 20, 1971), p. 47.

page 24, "In the discussion" quotation: From Paul R. Ehrlich and Richard W. Holm, "Patterns and Populations," *Science* 137 (August 31, 1962), p. 652; an article on "coevolution" is Paul R. Ehrlich and Peter H. Raven, "Butterflies and Plants: A Study in Coevolution," *Evolution* 18 (December 1964), pp. 586–608.

page 26, "Some commentators" quotation: From Paul R. Ehrlich, John P. Holdren, and Richard W. Holm, eds., *Man and the Ecosphere: Readings from Scientific American* (San Francisco: Freeman, 1971), p. 6.

page 26, Commoner *"some population-minded" quotation:* From Barry Commoner, *The Closing Circle* (New York: Knopf, 1971), p. 214.

page 26, Ehrlich *"I predict" quotation:* From "The Political and Social Responsibilities of Scientists" (cited above).

page 27, Ehrlich *"perhaps open" quotation:* From Constance Holden, "Ehrlich versus Commoner: An Environmental Fallout," *Science* 177 (July 21, 1972), p. 246. Commoner's denial of Stockholm chicanery is from the same article.

page 27, scientific disputes: On other public disputes following similar patterns, see Allan Mazur, "Disputes between Experts," *Minerva* 11 (April 1973), pp. 243–262; Dorothy Nelkin, "The Role of Experts in a Nuclear Siting Controversy," *Bulletin of the Atomic Scientists* 30 (November 1974), pp. 29–36.

pages 27–28, Science *magazine quotations:* From Constance Holden, "Ehrlich versus Commoner," *Science* 177 (July 21, 1972), pp. 245, 247.

page 28, Ehrlich-Commoner *articles:* In *Environment,* the two articles are entitled "Boardroom vs. Bedroom," *Environment* 14 (April 1972), pp. 23–26ff.; in the *Bulletin,* they are "Critique: One-Dimensional Ecology," and "Response," *Bulletin of the Atomic Scientists* 28 (May 1972), p. 17ff.

pages 28–29, Ehrlich-Commoner *letters:* Copies of the letters exchanged by Ehrlich, the *Environment* staff, and others were received by the author from an environment reporter who had received them from Ehrlich. The letters also indicate that: (1) Richard S. Lewis, editor of the *Bulletin* wrote Commoner on May 4, 1972, tracing a history of the *Bulletin-Environment* incident and concluding: "These circumstances suggest that my effort to accommodate you was exploited by *Environment* to 'scoop' our May issue. Whether the 'scoop' was intentional or not, I do not believe it enhances your side of the debate nor does it reflect any credit on *Environment*"; (2) On May 5, for the record, Ehrlich and Holdren wrote Lewis, noting that a cover letter accompanying the mimeographed preprints they had circulated had borne a "copyright notice" and a statement that it was a "preliminary copy," that it was "not for publication," and that the authors intended to publish a "version of it"; (3) On May 8, Ehrlich and Holdren wrote long letters to Daniel Kohl, publisher of *Environment,* and to the Science Advisory Board of *Environment.* Their letters prompted a letter from Harvard biologist John Edsall, as a member of the Science Advisory Board, to Sheldon Novick, editor of *Environment:* "If the facts there stated [in Ehrlich and Holdren's letter of May 8] are correct, it would appear that 'Environment' has been guilty of unethical conduct in printing the article by Ehrlich and Holdren in its April issue without permission"; (4) *En-*

vironment's defense is contained in letters from Novick on May 5 and Kohl on June 2 to Ehrlich and Holdren; Kohl's letter states in part: "Our attorneys judged the reproduction of your views [in the mimeographed draft] to be lacking a valid copyright, and therefore, to be in the public domain"; (5) Novick argues in a memorandum to the Science Advisory Board that Commoner told him he had informed the *Bulletin* that his article was being submitted to *Environment*; Lewis, editor of the *Bulletin*, says he was, on the contrary, taken by surprise.

page 31, Stuckey quotation: From William K. Stuckey, "John Bardeen," *Saturday Review of the Sciences* 1 (February 1973), p. 30.

page 32, four profiles of Mead: "Margaret Mead Today: Mother to the World," *Time* 93 (March 21, 1969), p. 74ff.; David Dempsey, "The Mead and Her Message," *New York Times Magazine*, April 26, 1970, p. 23ff.; Jeannie Sakol, "Remarkable Woman: Margaret Mead," *McCall's* 97 (June 1970), p. 81ff.; Winthrop Sargeant, "It's All Anthropology," *New Yorker* 37 (December 30, 1961), pp. 31–34ff.

page 33, Asimov "types" quotation: From Alfred Bester, "Isaac Asimov," *Publishers Weekly* 201 (April 17, 1972), p. 18.

page 34, de Camp quotation: From L. Sprague de Camp, "You Can't Beat Brains," *The Magazine of Fantasy and Science Fiction* 31 (October 1966), p. 33.

pages 34–35, Asimov poem: Isaac Asimov, "The Prime of Life," *The Magazine of Fantasy and Science Fiction* 31 (October 1966), p. 56; at Asimov's request, the words "mazel tov" have been substituted for "stars above" of the original version.

page 35, volume of science-news mail: From Victor Cohn, "Are We Really Telling the People About Science?" *Science* 148 (May 7, 1965), pp. 750–753.

page 36, 1972 study of thirty-nine visible scientists: See Rae Goodell, "The Visible Scientists," Ph.D. dissertation, Department of Communication, Stanford University, Stanford, Calif., 1975.

page 37, Davisson "overnight" quotation: Quoted by Lederberg from Göran Liljestrand, "The Prize in Physiology and Medicine," in H. Schuck et al., eds., *Nobel: The Man and His Prizes* (Stockholm: Nobel Foundation, 1950). Lederberg "The Nobel club" quotation is from a draft of an autobiographical essay provided to the author in an interview, February 16, 1973. On the public pressures surrounding receipt of the Nobel Prize, see also Harriet A. Zuckerman, *Scientific Elites: Nobel Laureates in the United States* (New York: Free Press-Macmillan, 1977).

GUERRILLA SCIENCE

page 39, "guerrilla science": Richard Levins of the University of Chicago used the title and term "guerilla science" (using the alternative spelling), independently of this book, in a speech prepared for a conference on "Scientists in the Public Interest," Alta, Utah, September 8, 1973; he defined "guerilla science" rather differently, as science which is "frankly dissident, partisan, and committed to radical social change."

page 39, political trends: Information to supplement this brief summary of the political trends in the United States that have produced today's visible scientists can be found in: Anne Hessing Cahn, "Eggheads and Warheads: Scientists and the ABM," Ph.D. dissertation, Science and Public Policy Program, Department of Political Science and Center for International Studies, M.I.T., Cambridge, Mass., 1971; Joseph Stefan Dupré and Sanford A. Lakoff, *Science and the Nation: Policy and Politics* (Englewood Cliffs, N.J.: Prentice-Hall, 1962); A. Hunter Dupree, *Science in the Federal Government: A History of Policies and Activities to 1940* (New York: Harper & Row, 1957); Robert Gilpin and Christopher Wright, eds., *Scientists and National Policy-Making* (New York: Columbia University Press, 1964); Daniel S. Greenberg, *The Politics of Pure Science* (New York: New American Library, 1967); Don K. Price, *The Scientific Estate* (New York: Oxford University Press, 1968); Dean Schooler, *Science, Scientists, and Public Policy* (New York: Free Press, 1971); Martin Sherwin, *A World Destroyed* (New York: Knopf, 1975); Eugene B. Skolnikoff, *Science, Technology, and American Foreign Policy* (Cambridge, Mass.: MIT Press, 1967); Alice Kimball Smith, *A Peril and a Hope: The Scientists' Movement in America 1945-47* (Cambridge, Mass.: MIT Press, 1971 edition); Donald A. Strickland, *Scientists in Politics: The Atomic Scientists Movement 1945-46* (Lafayette, Ind.: Purdue University Press, 1968); Ronald C. Tobey, *The American Ideology of National Science, 1919-1930* (Pittsburgh, Pa.: University of Pittsburgh Press, 1971); David Beckler, "The Changing Environment for National Science and Technology Policy" (paper delivered at the American Association for the Advancement of Science meeting, San Francisco, Calif., February 1974); Detlev W. Bronk, "Science Advice in the White House," *Science* 186 (October 11, 1974), pp. 116-121; G. B. Kistiakowsky, "Presidential Science Advising," *Science* 184 (April 5, 1974), pp. 38-42; Eugene B. Skolnikoff and Harvey Brooks, "Science Advice in the White House?" *Science* 187 (January 10, 1975), pp. 35-41.

page 40, Boulding quotation: From Kenneth E. Boulding, "Scientific Revelation," *Bulletin of the Atomic Scientists* 26 (September 1970), pp. 13-18.

page 40, scientists' noble impecunity: In the period before World War II, industry played a relatively large part in the funding of science and engineering research; see David Noble, *America by Design: Science, Technology and the Rise of Corporate Capitalism, 1880-1930* (New York: Knopf, 1977).

page 41, Bronowski quotation: From Jacob Bronowski, *Science and Human Values* (New York: Harper & Row, 1956), p. 75.

page 42, forty-seven federal agencies: From James L. McCamy, *Science and Public Administration* (University, Ala.: University of Alabama Press, 1960).

page 43, Pickering quotation: From William H. Pickering, "The Coming Change in Science and Technology," *Intellectual Digest* 4 (February 1974), p. 18.

page 43, McCabe quotation: From Charles Schwartz, Remarks presented at the Conference on Scientists in the Public Interest: The Role of Professional Societies, Alta, Utah, September 1973.

page 44, "arrogant piece" quotation: From Harvey Brooks, "The Practical Uses of Pure Research," *New York Times*, January 12, 1970., p. 78c.

page 44, Brenner quotation: From Sidney Brenner, interviewed by Charles

Weiner, May 21, 1975, Project on the Development of Recombinant DNA Guidelines, Oral History Collection, M.I.T. Archives, Cambridge, Mass. Quoted with permission.

page 45, "Science isn't" quotation: From Deborah Shapley, "White House Science: Hail and Farewell," Science 179 (March 30, 1973), p. 1311.

page 45, Etzioni and Nunn quotation: From Amitai Etzioni and Clyde Z. Nunn, "The Public Appreciation of Science in Contemporary America," Daedalus 103 (Summer 1974), p. 193.

page 45, survey of Californians: See Todd R. LaPorte and Daniel Metlay, "Technology Observed: Attitudes of a Wary Public," Science 188 (April 11, 1975), pp. 121–127.

page 46, Rockefeller quotation: From "Ford and Science: Honors, Warm Words," Science News 108 (September 27, 1975), p. 196.

page 47, Dupree quotation: From A. Hunter Dupree, Science in the Federal Government (New York: Harper & Row, 1957), p. 436.

page 47, "In 1966 a report" quotation: From Frank von Hippel and Joel Primack, "Public Interest Science," Science 177 (September 29, 1972), p. 1167. Other weaknesses of the science advisory system are raised in Meg Greenfield, "Science Goes to Washington," in W. R. Nelson, ed., The Politics of Science (New York: Oxford University Press, 1968); Robert K. Merton, "Insiders and Outsiders," American Journal of Sociology 78 (July 1972), pp. 9–47; Martin L. Perl, "The Scientific Advisory System," Science 173 (September 24, 1971), pp. 1211–1215; Joel Primack and Frank von Hippel, Advice and Dissent (New York: Basic Books, 1974).

page 48, Defense Department advisors: The figure of 86% Department of Defense ABM advisors receiving DoD funding is from Anne Hessing Cahn, "Eggheads and Warheads: Scientists and the ABM" (cited above).

page 48, Weinberg quotation: From Alvin M. Weinberg, Reflections on Big Science (Cambridge, Mass.: MIT Press, 1967), p. 69.

page 48, influential advisors: Estimates of the number of influential advisors are found in A. Leiserson, "Scientists and the Policy Process," American Political Science Review 59 (June 1965), pp. 408–416; and Christopher Wright, "Scientists and the Establishment of Science Affairs," in R. Gilpin and C. Wright, eds., Scientists and National Policy Making (New York: Columbia University Press, 1964), pp. 257–302.

page 48, "tame cat" quotation: From Philip H. Abelson, "Are the Tame Cats in Charge?" Saturday Review 49 (January 1, 1966), p. 100.

page 48, vacuum in scientific leadership: For an analysis similar to Watson's, see John Walsh, "Science Politics: An Invitation from the White House," Science 182 (October 26, 1973), pp. 365–368.

page 48, Gardner quotation: From John W. Gardner, "The Antileadership Vaccine," Annual Report of the Carnegie Corporation, New York, N.Y., 1965.

page 49, David analysis: From Edward E. David, Jr., "High Technology for a Livable World," remarks presented at symposium of U.S. presidential advisors, held at dedication ceremonies for Fairchild Electrical Engineering and Electronics Complex, M.I.T., Cambridge, Mass., October 4, 1975.

page 49, Kistiakowsky's analysis: From G. B. Kistiakowsky, "Presidential Science Advising," Science 184 (April 5, 1975), pp. 38–42.

page 50, Chicago physicist quotation: From Samuel S. Greenberg, *The Politics of Pure Science* (New York: New American Library, 1967), p. 117.

page 51, Kistiakowsky "Most of us" quotation: From Luther J. Carter, "After Cambodia and Kent: Academe Enters Congressional Politics," *Science* 168 (May 22, 1970), p. 955.

page 52, congressional fellows: See Benjamin S. Cooper and N. Richard Werthamer, "Two Physicists on Capitol Hill," *Physics Today* 28 (January 1975), pp. 63–66; Barbara J. Culliton, "Johnson Health Policy Fellows," *Science* 189 (September 19, 1975), pp. 977–980; Constance Holden, "Science Fellows in Washington," *Science* 189 (September 12, 1975), pp. 860–862.

page 52, Schwartz quotations: From Charles Schwartz, Remarks presented at the Conference on Scientists in the Public Interest: The Role of Professional Societies, Alta, Utah, September 1973. See also Charles Schwartz, "The Corporate Connection," *Bulletin of the Atomic Scientists* 31 (October 1975), pp. 15–19.

page 53, Ehrlich quotation: From Paul R. Ehrlich, *The Population Bomb* (New York: Ballantine, 1968), p. 175.

page 54, Seaborg "almost by accident" quotation: From Glenn Seaborg, untitled autobiographical essay, in Irving Stone, ed., *There Was Light* (Garden City, N.Y.: Doubleday, 1970), pp. 49–71.

page 54, Seaborg "In retrospect" quotation: From Glenn Seaborg, untitled autobiographical essay, in Irving Stone, ed., *There Was Light* (cited above), pp. 70–71.

page 57, "elder statesmen" quotation: From George A. W. Boehm, "AEC Gets a Different Kind of Scientist," *Fortune* 63 (April 1961), p. 158.

page 58, Boehm quotation: From George A. W. Boehm, "AEC Gets a Different Kind of Scientist," (cited above), pp. 160, 230.

page 58, Time quotations: From "Fallout over Seaborg," *Time* 97 (January 4, 1971), p. 49.

page 59, Seaborg "embarrassing" and "I really want" quotations: From "AAAS Presidency: Controversy Flares over Seaborg Candidacy," *Science* 170 (December 11, 1970), pp. 1177–1180.

page 59, Metzger quotations: From H. Peter Metzger, *The Atomic Establishment* (New York: Simon & Schuster, 1972), pp. 75–76.

page 60, Time "Paul Revere" quotation: Title of cover story in *Time* 95 (February 2, 1970), p. 58.

page 61, Revelle quotation: From Constance Holden, "Ehrlich versus Commoner," *Science* 177 (July 21, 1972), p. 245.

page 61, evolution of Commoner's views: See Barry Commoner, interviewed by D. Scott Peterson, April 24, 1973, Oral History Research Project, Department of History, Indiana University, Bloomington.

page 62, Commoner "What we need" quotation: From Barry Commoner, "The Fallout Problem," *Science* 127 (May 2, 1958), pp. 1025–1026.

page 62, history of SIPI: See "Scientists, Citizens and SIPI," SIPI Report 1, Fall 1970, p. 5ff.; and John Walsh, "Science Information: SIPI Expands, Puts New Emphasis on the Economy," *Science* 192 (April 9, 1976), pp. 122–124.

page 64, venereal disease article: Barry Commoner, "The Menace of Venereal Disease," *Science Illustrated* 1 (November 1946), p. 17ff.

page 65, report that won Newcomb Cleveland prize: Barry Commoner, et al., "The Proteins Synthesized in Tissue Infected with Tobacco Mosaic Virus," *Science* 118 (November 6, 1953), pp. 529–534.

page 66, Commoner "Biologists have" quotation: From Barry Commoner, "Failure of the Watson-Crick Theory As an Explanation of the Chemistry of Inheritance," *Nature* 220 (October 26, 1968), p. 349. On the subject of his holistic DNA theory, Commoner first published a general article, "In Defense of Biology," *Science* 133 (June 2, 1961), pp. 1745–1748. This was followed by an exchange of letters between Commoner and Isaac Asimov: "Modern Biology," *Science* 134 (October 6, 1961), p. 1020ff. Commoner then published three articles in *Nature*, the first of which had been rejected by *Science:* "Roles of Deoxyribonucleic Acid in Inheritance," *Nature* 202 (June 6, 1964), pp. 960–968; "Deoxyribonucleic Acid and the Molecular Basis of Self-Duplication," *Nature* 203 (August 1, 1964), pp. 486–491; "Failure of the Watson-Crick Theory," *Nature* 220 (October 26, 1968), pp. 334–340.

page 66, Crick quotation: From Francis Crick, "The Double Helix," *Nature* 248 (April 26, 1974), p. 768.

page 67, Commoner "What I'm doing" quotation: From Mary Catherine Bateson, *Our Own Metaphor* (New York: Knopf, 1972), p. 60.

page 68, Time *quotations:* From "Paul Revere of Ecology," *Time* 95 (February 2, 1970), p. 58.

page 68, Lang quotation: From Daniel Lang, "When Science Shoots the Works," *New York Times Book Review,* November 6, 1966, p. 58.

page 69, Time *quotation:* From "Paul Revere of Ecology" (cited above). Commoner says that the label "microbiologist" in the quotation is inaccurate, that he is a molecular biologist.

ALL IN THE FAMILY

page 70, chapter opening quotation: From Jacques Barzun, *Science: The Glorious Entertainment* (New York: Harper & Row, 1964), p. 137.

page 70, Chedd quotation: From Graham Chedd, "The Need to Understand," *New Scientist* 50 (June 24, 1971), p. 753.

page 72, Pauling "I had to tell" quotation: From William F. Fry, Jr., "What's New with You, Linus Pauling?" *The Humanist* 34 (November-December 1974), p. 17.

page 72, Pauling petition and Senate hearings: Pauling's test ban petition resulted from the audience's enthusiastic response to a speech he gave at Washington University, St. Louis, in 1957, as Pauling describes in his book *No More War!* (New York: Dodd, Mead, 1962). The petition ultimately collected over eleven thousand names in forty-nine countries. Barry Commoner says that he was one of the organizers of the petition, and one of the people whom Pauling protected in 1960 and 1961 by refusing to give the Senate Internal Security Subcommittee the names of those who helped with the petition. For further information on the Senate hearings, see Harry Kalven, Jr., "Congressional Testing of Linus Pauling," Parts I and II, *Bulletin of the Atomic Scientists* 16 (December 1960), pp. 383–390, and 17 (January 1961), pp. 12–19; "Pauling and the Senate Committee," *Science* 132 (October 14, 1960), p. 1001.

page 73, Nation *articles:* "The Pasternak Syndrome," *Nation* 191 (July 2, 1960), pp. 2–3; "The Efficacy of Protest," *Nation* 191 (October 29, 1960), p. 318.

page 73, U. S. News & World Report *quotation:* From "What a Senate Report Says about Efforts to Ban Bomb Tests," *U. S. News & World Report* 50 (April 10, 1961), p. 140. Other *U. S. News* articles on the subject are: "Congress Is Told How Pressure Grew to Ban Bomb Tests," *U. S. News & World Report* 42 (June 14, 1957), pp. 75–79; "Scientists and the Fall-Out Scare," *U. S. News & World Report* 42 (June 21, 1957), p. 52.

page 73, National Review *quotations:* From "The Collaborators," *National Review* 13 (July 17, 1962), p. 8; and "Are You Being Sued by Linus Pauling?" *National Review* 13 (September 25, 1962), p. 218.

page 73, Pauling's Nobel Peace Prize: A Nobel Peace Prize had not been awarded in 1962; Pauling's prize, announced in 1963, was called the 1962 Nobel Peace Prize.

page 73, Life *quotation:* From "A Weird Insult from Norway," *Life* 55 (October 25, 1963), p. 4.

page 73, Pauling "For many years" quotation: From "Pauling Welcomes Backing," *New York Times*, October 11, 1963, p. 26.

page 74, Silverman quotation: From "The Perils of Being Too Public," *Time* 87 (April 29, 1966), p. 80.

page 74, National Review *articles:* "Linus Pauling vs. National Review," *National Review* 17 (November 2, 1965), p. 962; "Linus Pauling—TKO," *National Review* 18 (May 3, 1966), p. 403.

page 75, Watson "Several members" quotation: From James D. Watson, *The Double Helix* (New York: New American Library, 1968), p. 79.

page 77, Wilkins quotation: From James D. Watson, *The Double Helix* (cited above), p. 21.

page 77, "felt sure" quotation: From an editor's note in Linus Pauling, "Fifty Years of Progress on Structural Chemistry and Molecular Biology," *Daedalus* 99 (Fall 1970), p. 1009.

page 78, Watson "Failure" quotation: From James D. Watson, *The Double Helix* (cited above), p. 79.

page 78, Pauling "I think I would" quotation: From Graham Chedd, "The Need to Understand," *New Scientist* 50 (June 24, 1971), p. 755. Pauling's mention of the alpha-helix refers to his study of a protein found in alpha keratin. Eleven years after his first research on the protein, Pauling, in collaboration with Robert B. Corey, worked out the protein's helical structure, which Pauling and Corey called an alpha-helix.

page 78, Science *quotation:* From Luther J. Carter, "Pauling Gets Medal of Science: Thaw between Scientists and the White House," *Science* 190 (October 3, 1975), p. 30. The National Medal of Science awards were instituted by Congress in 1959 and first awarded by President Kennedy in 1962. The prizes awarded by President Ford in September 1975 were for 1974.

page 79, Conservative Digest *quotation:* From a picture caption, Vol. 1 (December 1975), p. 28.

page 82, Pauliing "I was 69" quotation: From William F. Fry, Jr., "What's New with You, Linus Pauling?" *The Humanist* 34 (November-December 1974), p. 17.

page 82, Passmore quotation: From Reginald Passmore, "New Nostrum," *Nutrition Today* 6 (January-February 1971), p. 17.

page 82, vitamin C sales figures: From Jennifer Cross, "Vitamin Therapy," *San Francisco Bay Guardian* (October 18, 1972), pp. 4–5.

page 83, Life quotation: From Paul O'Neil, "The Vitamin C Mania," *Life* 71 (July 9, 1971).

page 83, Stare quotation: From Frederick J. Stare, "Not Quite Cricket," *Nutrition Today* 6 (January-February 1971), p. 18.

page 83, Mademoiselle article: "C: The Vitamin with Mystique," *Mademoiselle* 70 (November 1969), p. 189; other articles mentioned: Paul O'Neil, "The Vitamin C Mania," *Life* 71 (July 9, 1971), p. 55ff.; Sue Reilly, "The ABC's of Megavitamins," *Sky* 2 (August 1973), pp. 17–18.

page 83, Kent quotation: From Leticia Kent, "C, C: Dr. Linus Pauling Talks about Vitamin C and . . ." *Vogue* 157 (April 1, 1971), p. 130.

page 84, journal's rejection of Pauling's vitamin C papers: This information is from a conversation with Pauling and from Barbara J. Culliton, "Academy Turns Down a Pauling Paper," *Science* 177 (August 4, 1972), p. 409; John T. Edsall, "Linus Pauling and Vitamin C," *Science* 178 (November 17, 1972), p. 696; Linus Pauling, "Vitamin C," *Science* 177 (September 29, 1972), p. 1152. The three technical papers involved were Linus Pauling, "Evolution and the Need for Ascorbic Acid," *Proceedings of the National Academy of Sciences* 67 (December 15, 1970), pp. 1643–1648; Linus Pauling, "The Significance of the Evidence about Ascorbic Acid and the Common Cold," *Proceedings of the National Academy of Sciences* 68 (November 1971), pp. 2678–2681; Ewan Cameron and Linus Pauling, "Ascorbic Acid and Glycosaminoglycans," *Oncology* 27 (1973), pp. 181–192.

page 85, Cousins quotation: From Norman Cousins, "Linus Pauling and the Vitamin Controversy," *Saturday Review* 54 (May 15, 1971), pp. 37, 44.

page 85, Enloe quotation: From Cortez F. Enloe, Jr., "The Virtue of Theory," *Nutrition Today* 6 (January-February 1971), p. 21.

page 86, Chedd quotations: From Graham Chedd, "The Need to Understand," *New Scientist* 50 (June 24, 1971), pp. 753, 755.

page 87, Scientific American quotations: From "Linus Pauling," *Scientific American* 145 (November 1931), p. 293.

page 87, Saturday Evening Post quotation: From M. Grosser, "Linus Pauling: Molecular Artist," *Saturday Evening Post* 243 (Fall 1971), p. 14.

page 87, Daedalus article: Linus Pauling, "Fifty Years of Progress on Structural Chemistry and Molecular Biology," *Daedalus* 99 (Fall 1970), pp. 988–1014.

page 87, orthomolecular psychiatry: See Linus Pauling, "Orthomolecular Psychiatry," *Science* 160 (April 19, 1968), pp. 265–271.

page 87, Barber quotation: From Bernard Barber, "Resistance by Scientists to Scientific Discovery," *Science* 134 (September 1, 1961), p. 601.

page 88, Rosenfeld quotation: From Albert Rosenfeld, "The 'Factifuging' Syndrome," *Saturday Review/World* 1 (January 26, 1974), p. 65.

page 88, social system of science: See Jonathan R. Cole and Stephen Cole, *Social Stratification in Science* (Chicago: University of Chicago Press, 1973); Bernice T. Eiduson, *Scientists: Their Psychological World* (New York: Basic Books, 1962); Barney G. Glaser, *Organizational Scientists:*

Their Professional Careers (Indianapolis: Bobbs-Merrill, 1964); W. O. Hagstrom, *The Scientific Community* (New York: Basic Books, 1965); Norman Kaplan, "The Sociology of Science," in Robert E. Faris, ed., *Handbook of Modern Sociology* (Chicago: Rand McNally, 1964), pp. 852–881; Robert K. Merton, *Social Theory and Social Structure* (New York: Free Press, 1968); N. W. Storer, *The Social System of Science* (New York: Holt, Rinehart and Winston, 1966); John Ziman, *Public Knowledge: The Social Dimension of Science* (Cambridge, England: Cambridge University Press, 1968). The phrase "ethos of science" is from Merton (cited above). On various manifestations of the reward system, see Stephen Cole and Jonathan R. Cole, "Scientific Output and Recognition," *American Sociological Review* 32 (June 1967), pp. 377–390; Diana Crane, "Scientists at Major and Minor Universities: A Study of Productivity and Recognition," *American Sociological Review* 30 (October 1965), pp. 699–714; Robert K. Merton, "Priorities in Scientific Discovery," *American Sociological Review* 22 (December 1957), pp. 635–659; Robert K. Merton, "The Matthew Effect in Science," *Science* 159 (January 1968), pp. 56–63; F. Reif, "The Competitive World of the Pure Scientist," *Science* 134 (December 1961), pp. 1957–1962; Harriet A. Zuckerman and Robert K. Merton, "Patterns of Evaluation in Science: Institutionalization, Structure and Functions of the Referee System," *Minerva* 9 (January 1971), pp. 66–100; Harriet A. Zuckerman, "Stratification in American Science," *Sociological Inquiry* 40 (Spring 1970), pp. 235–257; Harriet A. Zuckerman, "Nobel Laureates in Science: Patterns of Productivity, Collaboration, and Authorship," *American Sociological Review* 32 (June 1967), pp. 391–403.

page 89, Barzun quotation: From Jacques Barzun, *Science: The Glorious Entertainment* (New York: Harper & Row, 1964), p. 120.

page 89, "once so despaired" quotation: From Mitchell Wilson, "On Being a Scientist," *Atlantic* 226 (September 1970), p. 102.

page 90, Klaw quotation: From Spencer Klaw, *The New Brahmins: Scientific Life in America* (New York: Morrow, 1968), p. 12.

page 90, psychological characteristics of scientists: See Francis Bello, "The Young Scientists," in *Fortune* magazine, eds., *The Mighty Force of Research* (New York: McGraw-Hill, 1956), pp, 21–39; Bernice T. Eiduson, *Scientists: Their Psychological World* (New York: Basic Books, 1962); Spencer Klaw, *The New Brahmins* (cited above); R. H. Knapp and H. B. Goodrich, *Origins of American Scientists* (New York: Russell and Russell, 1952); David C. McClelland, "On the Psychodynamics of Creative Physical Scientists," in H. E. Gruber, G. Terrell, and M. Werthumer, eds., *Contemporary Approaches to Creative Thinking* (New York: Atherton, 1962), pp. 141–174; Anne Roe, *The Making of a Scientist* (New York: Dodd, Mead, 1953); Calvin W. Taylor and Frank Barron, eds., *Scientific Creativity: Its Recognition and Development* (New York: Wiley, 1963); Stephen S. Visher, *Scientists Starred 1903–1943* (Baltimore, Md.: Johns Hopkins Press, 1947).

page 91, Abelson quotation: From Philip H. Abelson, "Are the Tame Cats in Charge?" *Saturday Review* 49 (January 1, 1966), p. 103.

page 91, Kornberg quotation: From Norman Melnick, "Scientist Speaks Out for 'Truth,'" *San Francisco Sunday Examiner and Chronicle* (December 19, 1971), p. 22.

page 91, conflicting values in science: See Robert K. Merton, "The Ambivalence of Scientists," in N. Kaplan, ed., *Science and Society* (Chicago: Rand McNally, 1965), pp. 112–132; Harriet A. Zuckerman, "Patterns of Name Ordering among Authors of Scientific Papers: A Study of Social Symbolism and Its Ambiguity," *American Journal of Sociology* 74 (November 1968), pp. 276–291. On changing values in science, see André F. Cournand and Harriet A. Zuckerman, "The Code of Science Analysis and Some Reflections on the Future," *Studium Generale* 23 (1970), pp. 941–962.

page 93, Lederberg quotation: From a draft of an autobiographical essay provided to the author in interview, February 16, 1973.

page 97, Tobey's thesis: Ronald C. Tobey, *The American Ideology of National Science 1919–1930* (Pittsburgh, Pa.: University of Pittsburgh Press, 1971).

WHAT PRICE SUCCESS?

page 102, Asimov quotation: From Isaac Asimov, "Academe and I," *The Magazine of Fantasy and Science Fiction* 42 (May 1972), pp. 141–142.

page 104, "lined and triple-lined" quotation: From Richard Todd, " 'Walden Two': Three? Many More?" *New York Times Magazine,* March 15, 1970, p. 119.

page 104, Mead "I was so" quotation: From Margaret Mead, *Blackberry Winter* (New York: Morrow, 1972), p. 160.

page 105, sequence in Science *magazine:* (1) article by Paul R. Ehrlich and Richard W. Holm, "Patterns and Populations," *Science* 137 (August 31, 1962), pp. 652–657; (2) letter by Dean Amadon, "Population Biology," *Science* 138 (November 9, 1962), pp. 733–734, with Ehrlich and Holm's one-sentence reply; (3) letter criticizing Ehrlich and Holm's "flippant reply" by Grady L. Webster, "Population Biology," *Science* 139 (January 18, 1963), p. 236ff., with longer reply by Ehrlich and Holm.

page 105, "Old India" quotations: From Paul R. Ehrlich, *The Population Bomb* (New York: Ballantine, 1968), pp. 15–16; and from the 1971 edition, pp. 1–2.

page 107, The New York Times Magazine quotation: From Berkeley Rice, "Skinner Agrees He Is the Most Important Influence in Psychology," *New York Times Magazine,* March 17, 1968, p. 27.

page 107, Psychology Today quotation: From Elizabeth Hall, "Will Success Spoil B. F. Skinner?" *Psychology Today* 6 (November 1972), p. 68.

page 107, Platt quotation From Elizabeth Hall (cited above), p. 68.

page 107, Szasz quotation: From Thomas S. Szasz, "A Critique of Skinner's Behaviorism," *The Humanist* 35 (March-April 1975), p. 31.

page 108, "rose at 6 a.m." quotation: From Berkeley Rice, "Skinner Agrees He Is the Most Important Influence in Psychology" (cited above), p. 108.

page 108, Skinner's cumulative record: From Richard I. Evans, *B. F. Skinner: The Man and His Ideas* (New York: Dutton, 1968).

page 108, Skinner "I treat myself" quotation: From Berkeley Rice, "Skinner Agrees He Is the Most Important Influence in Psychology" (cited above), p. 112.

page 108, Time *"The staffers" quotation:* From "Skinner's Utopia: Panacea, or Path to Hell?" *Time* 98 (September 20, 1971), p. 52.

page 109, Time *"Dr. Skinner" quotation:* From "Pigeons and People," *Time* 55 (June 19, 1950), p. 73.

page 109, Rice quotation: From Berkeley Rice, "Skinner Agrees He Is the Most Important Influence in Psychology" (cited above), p. 90.

page 110, Skinner "It would be" quotation: From Richard Todd, " 'Walden Two': Three? Many More?" *New York Times Magazine,* March 15, 1970, p. 120.

page 110, Keller quotation: From Fred S. Keller, "Psychology at Harvard (1926–1931)," in P. B. Dews, ed., *Festshrift for B. F. Skinner* (New York: Irvington, 1970), p. 35.

page 111, Herrnstein quotation: From Berkeley Rice, "Skinner Agrees He Is the Most Important Influence in Psychology" (cited above), p. 95.

page 111, Ladies' Home Journal *article:* B. F. Skinner, "Baby in a Box," *Ladies' Home Journal* 62 (October 1974), pp. 30–31.

page 112, Deborah Skinner identity crisis: From "Skinner's Utopia: Panacea, or Path to Hell?" (cited above), p. 47ff.

page 113, Frost quotation: From Lawrence Thompson, ed., *Selected Letters of Robert Frost* (New York: Holt, Rinehart and Winston, 1964), p. 327; also in B. F. Skinner, *Particulars of My Life* (New York: Knopf, 1976).

page 113, Skinner "the results," "hack," "Bohemian," and "I was" quotations: From B. F. Skinner, untitled autobiographical essay, in E. G. Boring and G. Lindzey, eds., *A History of Psychology in Autobiography,* Vol. V (New York: Appleton-Century-Crofts, 1967), pp. 394, 395.

page 114, Skinner "To my surprise" quotation: From Richard Todd, " 'Walden Two': Three? Many More?" (cited above), p. 24.

page 116, Skinner "I am convinced" quotation: From Richard I. Evans, *B. F. Skinner: The Man and His Ideas* (New York: Dutton, 1968), p. 106.

page 117, Saturday Review *quotations:* From Walter Arnold, Review of B. F. Skinner, *Beyond Freedom and Dignity, Saturday Review* 54 (October 9, 1971), p. 47.

page 117, Atlantic quotation: From George Kateb, "Toward a Wordless World," *Atlantic* 228 (October 1971), p. 122.

page 119, Skinner "what I think" quotation: From Richard I. Evans, *B. F. Skinner: The Man and His Ideas* (cited above), p. 111.

WHAT'S A NICE SCIENTIST DOING IN A
PLACE LIKE THE PRESS?

page 120, chapter opening quotation: The inadvertent claim that vitamin C may grant immortality can be assumed to be the *Midnight* headline writer's, not Pauling's.

page 120, Salk quotation: From Richard Carter, *Breakthrough: The Saga of Jonas Salk* (New York: Trident, 1966), p. 292.

page 120, Look reporter quotation: From Richard Carter (cited above), p. 166.

page 121, March of Dimes life insurance: From Nicholas Wade, "Salk Worth $2.4 Million to March of Dimes," *Science* 183 (March 8, 1974), p. 937.

page 121, Einstein quotation: From Ronald W. Clark, *Einstein: The Life and Times* (New York: World Publishing, 1971).

page 122, Abelson quotation: From Philip H. Abelson, "Science Reporting," *Science* 139 (January 18, 1963), p. 177.

page 122, Drake quotation: From Donald C. Drake, "A Science Writer Looks at the American Newspaper," *AAAS Bulletin* 17 (April 1972), p. 3.

page 122, sensationalism in science news: For the amount of sensationalism, see George Gerbner, "The Press and the Dialogue in Education," *Journalism Monographs*, No. 5 (September 1967); Glynn L. Wood, "A Scientific Convention as Source of Popular Information," in Studies of Innovation and of Communication to the Public, Vol. III of "Studies in the Utilization of Behavioral Science," Institute for Communication Research, Stanford University, Stanford, Calif., 1962; Phillip J. Tichenor et al., "Mass Communication Systems and Communication Accuracy in Science News Reporting," *Journalism Quarterly* 47 (Winter 1970), pp. 673–683; James W. Tankard, Jr., and Michael Ryan, "News Source Perceptions of the Accuracy of Science Coverage," *Journalism Quarterly* 51 (Summer 1974), pp. 219–225.

For the legitimate role of sensationalism in science news, see Victor Cohn, "Are We Really Telling the People about Science?" *Science* 148 (May 7, 1965), pp. 750–753; Muriel Davidson, "Viewer, Heal Thyself!" *TV Guide* 21 (July 21, 1973), pp. 21–24; E. G. Sherburne, Jr., "Science on Television: A Challenge to Creativity," *Journalism Quarterly* 40 (Summer 1963), p. 300; John Troan, "Science Reporting, Today and Tomorrow," *Science* 131 (April 22, 1960), pp. 1193–1196. On the need for more qualification in science news, see "Judging What to Use from the Wires," panel presented at a seminar on Medical Science in the News, Chicago Press Club, November 1965; Clay Schoenfeld, "How to Write about Science," *Writer* 77 (February 1964), pp. 22–25.

page 122, Clark quotation: From Marguerite Clark, "Today: Medicine; Tomorrow: RNA," National Association of Science Writers *Newsletter* 8 (September 1960), p. 22.

page 122, "I'll take gee-whiz" quotation: From Stan Wiggins, "Ehrlich and Commoner," *Bulletin of the Atomic Scientists* 28 (September 1972), p. 2.

page 122, psoriasis research story: In Michael Ryan and James W. Tankard, Jr., untitled article in "Science in the Newspaper," American Association for the Advancement of Science, Washington, D.C. 1974, p. 28.

page 123, Drake quotation: Donald C. Drake, "A Science Writer Looks at the American Newspaper" (cited above), pp. 3–4; repercussions of Drake's article are discussed in Dan Rottenberg, "Biting the Hand . . . Reviewing Journalism Is a Hazardous Occupation," *Chicago Journalism Review* 5 (June 1972), pp. 3–4.

page 123, need for historical perspective: See J. J. Fortman, G. G. Hess, and D. J. Karl, "Science and the Popular Press," *Chemistry* 44 (October 1971), pp. 20–21; Polykarp Kusch, "The World of Science and the Scientist's World," *Bulletin of the Atomic Scientists* 24 (August 1968), pp. 38–43.

page 123, Krieghbaum remarks: From Hillier Krieghbaum, *Science and the Mass Media* (New York: New York University Press, 1967).

page 123, instant research: See Luther P. Jackson, Jr., "Communicating Research on the Poor," *Negro History Bulletin* 29 (April 1966), p. 151ff.

page 124, science news accuracy: For a more detailed discussion of the research on science news accuracy, suggesting that science news is no more and no less accurate than other news, see the author's Ph.D. dissertation, "The Visible Scientists," Department of Communication, Stanford University, Stanford, Calif., 1975.

page 124, "you guys" quotation: From William L. Rivers, *Finding Facts: Interviewing, Observing, Using Reference Sources* (Englewood Cliffs, N. J.: Prentice-Hall, 1975).

page 125, Thistle quotation: From M. W. Thistle, "Popularizing Science," *Science* 127 (April 25, 1958), p. 952.

page 125, Tannenbaum thesis: See Percy H. Tannenbaum, "Communication of Science Information," *Science* 140 (May 10, 1963), pp. 579–583; see also Leo Bogart, "Changing News Interests and the News Media," *Public Opinion Quarterly* 32 (Winter 1968–69), pp. 560–574; Kenneth G. Johnson, "Dimensions of Judgment of Science News Stories," *Journalism Quarterly* 40 (Summer 1963), p. 315; Joye Patterson, Laurel Booth, and Russell Smith, "Who Reads about Science?" *Journalism Quarterly* 46 (Autumn 1969), pp. 599–602.

page 126, training of science writers: See "Report on Conference on the Role of Schools of Journalism in the Professional Training of Science Writers," *Science News Letter* 79 (June 17, 1961), p. 379; Pierre C. Fraley, "The Education and Training of Science Writers," *Journalism Quarterly* 40 (Summer 1963), p. 323; Sharon M. Friedman, "Changes in Science Writing since 1965 and Their Relation to Shifting Public Attitudes toward Science," Master's thesis, Department of Journalism, Pennsylvania State University, University Park, 1974; Lee Z. Johnson, "Status and Attitudes of Science Writers," *Journalism Quarterly* 34 (Spring 1957), p. 247; Hillier Krieghbaum, *Science and the Mass Media* (New York: New York University Press, 1967); Michael Ryan and Sharon L. Dunwoody, "Academic and Professional Training Patterns of Science Writers," *Journalism Quarterly* 52 (Summer 1975), p. 239ff.; William E. Small, "Science Writer Survey," National Association of Science Writers *Newsletter* 11 (December 1963), pp. 11–15; Phillip J. Tichenor et al., "Mass Communication Systems and Communication Accuracy in Science News Reporting," *Journalism Quarterly* 47 (Winter 1970), pp. 673–683. Friedman, as well as Ryan and Dunwoody, found a trend toward more science training among science writers.

page 126, Carey quotation: From Frank Carey, "A Quarter Century of Science Reporting," *Nieman Reports* 20 (June 1966); on the need for on-the-job training, see also D. M. Gates and J. M. Parker, "Science News Writing: Seminar at Colorado State University," *Science* 133 (January 20, 1961), pp. 211–214.

page 126, Time quotation: From "Science of Reporting: Career of *New York Times's* Science Editor," *Time* 82 (December 27, 1963), p. 32.

page 127, 0.01 percent estimate: From M. W. Thistle, "Popularizing Science," *Science* 127 (April 25, 1958), pp. 951–955; the estimate takes into account not only media constriction but also language, psychological, and secrecy barriers. See also William J. Paisley, "The Flow of (Behavioral) Science Information: A Review of the Research Literature," Institute for Communication Research, Stanford University, Stanford, Calif., 1965.

page 127, Perlman quotation: From David Perlman, remarks delivered at seminar on "Public Understanding of Science," American Association for the Advancement of Science, San Francisco, Calif., February 1974.

page 127, three stages of science writing: See "Science Reporting Has Grown Out of Its 'Gee Whiz' Phase," *Editor and Publisher* 103 (September 12, 1970), p. 20; Victor Cohn, "Are We Really Telling the People about Science?" *Science* 148 (May 7, 1965), pp. 750–753; Lawrence Lessing, "Science Journalism: The Coming of Age," *Bulletin of the Atomic Scientists* 19 (December 1963), p. 23.

page 127, Carey remarks: From Frank Carey, "A Quarter Century of Science Reporting," *Nieman Reports* (cited above), pp. 7–10.

page 128, Skardon quotation: From James A. Skardon, "The Apollo Story: What the Watchdogs Missed," *Columbia Journalism Review* 6 (Fall 1967), pp. 11–15 and (Winter 1967–68), pp. 34–39.

page 128, increase in unfavorable coverage of science: See Sharon M. Friedman, "Changes in Science Writing since 1965" (cited above).

page 128, Cohn quotation: From Victor Cohn, "Are We Really Telling the People about Science?" (cited above), p. 753.

page 128, Krieghbaum quotation: From Hillier Krieghbaum, *Science and the Mass Media* (cited above), p. 160.

page 129, Graham quotation: From Frank Graham, Jr., *Since Silent Spring* (Boston: Houghton Mifflin, 1970), pp. 165–166.

page 129, Project Argus story: From William L. Rivers, *The Opinionmakers* (Boston: Beacon Press, 1965), pp. 78–81.

page 130, NASA press restrictions: See David W. Burkett, *Writing Science News for the Mass Media*, 2nd ed. (Houston, Tex.: Gulf Publishing, 1973); Bruce Jon Cole, "Science Conflict: A Content Analysis of Four Major Metropolitan Newspapers, 1951, 1961, 1971," Master's thesis, School of Journalism and Mass Communications, University of Minnesota, Minneapolis, 1974; Hillier Krieghbaum, *Science and the Mass Media* (cited above).

page 130, Ingelfinger rule: See Stuart S. Blume, *Toward a Political Sociology of Science* (New York: Free Press, 1974), p. 226; Barbara J. Culliton, "Dual Publication: 'Ingelfinger Rule' Debated by Scientists and Press," *Science* 176 (June 30, 1972), pp. 1403–1405.

page 131, Lear quotation: From John Lear, "When Is a New Idea Fit to Print?" *Saturday Review* 45 (April 7, 1962), p. 45.

page 131, Cohn "Neither journalist" quotation: From Victor Cohn, "Are We Really Telling the People about Science?" (cited above), p. 752.

page 131, increasing volume of technical literature: See Derek J. Price, *Little Science, Big Science* (New York: Columbia University Press, 1963).

page 131, Cohn "guilty" quotation: Victor Cohn, "Are We Really Telling the People about Science?" (cited above).

page 132, Sprat quotation: From Meta R. Emberger and Marian R. Hall, *Science Writing* (New York: Harcourt, Brace, 1955), pp. 129–130.

page 133, "After fifteen years" quotation: From Orest Dubas and Lisa

Martel, *Media Impact*, Vol. 1, Interim Report, Ministry of State, Science and Technology, Ottawa, Canada, October 1973, p. 36.

page 134, Blakeslee anecdote: From David W. Burkett, *Writing Science News for the Mass Media* (cited above), p. 27.

page 134, "corridor consultation" quotation: From Walter Sullivan, "Writing Science for the Public," *Physics Today* 23 (August 1970), pp. 51–53.

page 135, "pack reporting" quotation: From Theodore F. Koop, "The Communication of Science: Tainted by Experience," *Federation of American Societies for Experimental Biology Proceedings* 32 (April 1973), pp. 1442–1444.

page 135, characteristics of science news consumers: See William J. Paisley, "The Flow of (Behavioral) Science Information: A Review of the Research Literature," Institute for Communication Research, Stanford University, Stanford, Calif., 1965; Matilda B. Rees and William J. Paisley, "Social and Psychological Predictors of Information-Seeking and Media Use," Institute for Communication Research, Stanford University, Stanford, Calif., 1967; Wilbur Schramm and Serena Wade, "Knowledge and the Public Mind," Institute for Communication Research, Stanford University, Stanford, Calif., 1967; "The Public Impact of Science in the Mass Media," Survey Research Center, University of Michigan, Ann Arbor, 1958.

page 135, readers' psychological mechanisms: See Raymond A. Bauer, "The Obstinate Audience," *American Psychologist* 19 (May 1964), pp. 319–328; Kenneth E. Boulding, *The Image: Knowledge of Life and Society* (Ann Arbor: University of Michigan Press, 1956); Jerome S. Bruner, "Social Psychology and Perception," in E. E. Maccoby, T. M. Newcomb, and E. L. Hartley, eds., *Readings in Social Psychology* (New York: Henry Holt, 1958), pp. 85–94; Albert H. Hastorf and Hadley Cantril, "They Say a Game: A Case Study," *Journal of Abnormal and Social Psychology* 49 (1954), pp. 129–134; Daniel Katz, "Psychological Barriers to Communication," in W. Schramm, ed., *Mass Communications* (Urbana: University of Illinois Press, 1949), pp. 275–289; Walter Lippmann, *Public Opinion*, 2nd ed., (New York: Free Press, 1965).

page 135, University of Missouri study: "pScience: Science in Society," School of Journalism, University of Missouri, Columbia, December 1973. Other studies showing the importance of relevance and practicality in science news include: Amitai Etzioni and Clyde Z. Nunn, "The Public Appreciation of Science in Contemporary America," *Daedalus* 103 (Summer 1974), pp. 101–205; "Science Indicators 1972," Report of the National Science Board, National Science Foundation, 1973; Wilbur Schramm and Serena Wade, "Knowledge and the Public Mind" (cited above); James W. Swinehart and Jack M. McLeod, "News about Science: Channels, Audiences, and Effects," *Public Opinion Quarterly* 24 (Winter 1961), pp. 583–589; "The Public Impact of Science in the Mass Media (cited above); "Satellites, Science, and the Public," Survey Research Center, University of Michigan, Ann Arbor, 1959; Serena Wade and Wilbur Schramm, "The Mass Media as Sources of Public Affairs, Science and Health Knowledge," *Public Opinion Quarterly* 33 (Summer 1969), pp. 197–209.

page 141, "If a non-anthropological" quotation: From Malcolm C. Webb, "The Culture Concept and Cultural Change in the Work of Margaret

Mead," *Proceedings of the Louisiana Academy of Sciences* 31 (December 31, 1968), p. 148.

page 142, Worsley quotation: From Peter M. Worsley, "Margaret Mead: Science or Science Fiction? Reflections of a British Anthropologist," *Science and Society* 21 (Spring 1957), p. 122.

page 142, Time quotation: From "Margaret Mead Today: Mother to the World," *Time* 93 (March 21, 1969), p. 74.

page 143, Mead Blackberry Winter *quotations:* From Margaret Mead, *Blackberry Winter: My Earlier Years* (New York: Morrow, 1972), pp. 81, 109. .

page 144, Mead's bibliography: Joan Gordan, ed., *Margaret Mead: The Complete Bibliography 1925–1975* (New York: Mouton, 1976).

page 144, "of being able" quotation: From Allyn Moss, *Shaping a New World: Margaret Mead* (Chicago: Encyclopaedia Britannica Press, 1963), p. 165.

page 145, "Meaty Meadisms" quotation: "Meaty Meadisms About America," *Life* 47 (September 14, 1959), p. 147.

page 145, Mead "Well, it isn't" quotation: From David Dempsey, "The Mead and Her Message," *New York Times Magazine*, April 26, 1970, p. 82.

page 145, Mead "Motherhood is like" quotation: From "Meaty Meadisms About America" (cited above), p. 147.

page 145, Mead "Different cultures" quotation: From Irene Neves, "We Must Learn to See What's Really New," *Life* 65 (August 23, 1968).

page 145, Mead "Women are much fiercer" quotation: From Jeannie Sakol, "Remarkable Woman: Margaret Mead," *McCall's* 97 (June 1970), p. 127.

page 146, Mead "For the first time" quotation: From "Margaret Mead Today" (cited above), p. 74.

page 146, Mead "Fathers are spending" quotation: From "Meaty Meadisms About America" (cited above), p 147

page 146, Mead "Of course I believe" quotation: From "People, Etc.," *Boston Sunday Globe Magazine*, October 13, 1974, p. 41.

page 146, Dempsey quotation: From David Dempsey, "The Mead and Her Message" (cited above), p. 23.

page 146, decision to go to Samoa: Mead describes the decision in *Blackberry Winter* (cited above), pp. 126–129.

page 147, Mead "I had what amounted" quotation: From Margaret Mead, *Blackberry Winter* (cited above), p. 224.

page 148, Time *quotation:* From "Margaret Mead Today" (cited above), p. 77.

page 149, Worsley "The professional" quotation: From Peter Worsley, "Margaret Mead: Science or Science Fiction?" (cited above), p. 122.

page 149, Mead's critics: Examples of critical reviews of Mead's work are: Thomas Gladwin, Review of Margaret Mead and Martha Wolfenstein, *Childhood in Contemporary Cultures*, in *American Anthropologist* 58 (1956), pp. 764–767; F. Kluckhohn and C. Kluckhohn, Review of Margaret Mead, *And Keep Your Powder Dry*, in *American Anthropologist* 45 (1943), pp. 622–624; S. F. Nadel, Review of Margaret Mead, *Male and Female*, in *American Anthropologist* 52 (1950), pp. 419–420.

page 149, "Though she tackles," quotation: From Peter Worsley, "Margaret Mead: Science or Science Fiction?" (cited above), pp. 122–123.

page 150, Worsley "The aggrieved specialist" quotation: From Peter Worsley, "Margaret Mead: Science or Science Fiction?" (cited above), p. 125.

page 151, Haring quotation: From Douglas Haring, Review of Margaret Mead, *Continuities in Cultural Evolution,* in *Current Anthropology* 7 (February 1966), p. 72.

page 151, Tax quotation: From Sol Tax, Review of Margaret Mead, *Continuities in Cultural Evolution,* in *Current Anthropology* 7 (February 1966), p. 76.

page 152, McCall's quotations: From Jeannie Sakol, "Remarkable Woman: Margaret Mead" (cited above), p. 81.

page 152, Neves quotation: From Irene Neves, "We Must Learn to See What's Really New" (cited above), p. 30.

page 152, Time quotation: From "Margaret Mead Today" (cited above), p. 74.

page 153, "everybody's grandmother" quotation: From Caroline Bird, "Everybody's Grandmother," *Saturday Review* 55 (November 25, 1972), p. 64.

page 153, "mother to the world" quotation: From "Margaret Mead Today" (cited above), p. 74.

page 153, "She could pass" quotation: From David Dempsey, "The Mead and Her Message" (cited above), p. 74.

page 153, "Although the ankle" quotation: From David Dempsey, "The Mead and Her Message" (cited above), p. 77.

page 155, Kubota article: Irene Kubota, "An Interview with Margaret Mead," *Redbook* 143 (August 1974), p. 31ff.

page 155, "The doses" quotation: From Winthrop Sargeant, "It's All Anthropology," *New Yorker* 37 (December 30, 1961), p. 31.

page 155, Mead "When I began" quotation: From Margaret Mead, *Blackberry Winter* (cited above), p. 289.

page 156, Saturday Review quotations: From Caroline Bird, "Everybody's Grandmother" (cited above), pp. 64, 66.

page 157, Mead "political success" and "In a curious way" quotations: From Margaret Mead, *Blackberry Winter* (cited above), pp. 10–11.

page 158, Man and Nature Lectures: These were later published in *Culture and Commitment: A Study of the Generation Gap* (Garden City, N. Y.: National History Press/Doubleday, 1970).

page 158, Bird "I have to confess" quotation: From Caroline Bird, "Everybody's Grandmother" (cited above), p. 64.

page 158, Harris "the courage" quotation: From "Margaret Mead Today" (cited above), p. 77.

page 160, New Yorker profile: Winthrop Sargeant, "It's All Anthropology" (cited above); "mediocre art" reference is on p. 33.

page 161, "Early this year" quotation: From David Dempsey, "The Mead and Her Message" (cited above), p. 23.

THE POPULAR CONNECTION

page 163, chapter opening quotation: From Carl Sagan, *Other Worlds,* produced by Jerome Agel (New York: Bantam, 1975), p. 145.

page 163, New Scientist *quotation:* From Ian Ridpath, "A Man Whose Time Has Come," *New Scientist* 63 (July 4, 1974), p. 36.

page 164, Sagan *"popularity of astrology" quotation:* From Carl Sagan, *Other Worlds* (cited above), p. 123.

page 164, Sagan *"If the eons" quotation:* From Stuart Baur, "Kneedeep in the Cosmic Overwhelm with Carl Sagan," *New York* 8 (September 1, 1975), p. 28; the same analogy is used in *Other Worlds* (cited above), p. 12.

page 165, Sagan's *license plates:* Sagan lives in Ithaca, New York, where Cornell University is located, but often spends part of the year in California, where he gets California vanity license plates.

page 166, Sagan's *childhood experiences:* Similar childhood memories, reported in more detail, can be found in a two-part biographical profile of Sagan: Henry S. F. Cooper, "A Resonance with Something Alive," *New Yorker* 52 (June 21, 1976), p. 39ff., and 52 (June 28, 1976), p. 30ff.

page 169, Sagan *"Nothing much" quotation:* From Stuart Baur, "Kneedeep in the Cosmic Overwhelm with Carl Sagan" (cited above), p. 31.

page 170, CETI *book:* Carl Sagan, ed., *Communication with Extraterrestrial Intelligence (CETI)* (Cambridge, Mass.: MIT Press, 1973).

page 170, Ridpath *quotation:* From Ian Ridpath, "A Man Whose Time Has Come" (cited above), p. 37.

page 171, Baur *quotations:* From Stuart Baur, "Kneedeep in the Cosmic Overwhelm with Carl Sagan" (cited above), pp. 28, 32. Baur indicates that Sagan's first appearance on the Johnny Carson show was in 1970; Sagan recalls it as 1972.

page 172, New York Times *article:* "Mars' Dark Spots Held to Be Lava," *New York Times,* December 29, 1956, p. 6.

page 174, Time *feature article:* "Is There Life on Mars — or Beyond?" *Time* 98 (December 13, 1971), pp. 50–52ff.

page 174, Time *list of leaders:* "200 Faces for the Future," *Time* 104 (July 15, 1974), pp. 35–40ff.

page 174, Saturday Review *list:* Susan Schiefelbein, "Future Promise," *Saturday Review/World* 2 (December 14, 1974), pp. 95–97.

page 176, Sagan *"like a dandelion" quotation:* From "The Planets," NOVA series, WGBH television, Boston, Mass., January 25, 1976.

HYPING SCIENCE

page 178, Metzger *quotation:* From H. Peter Metzger, Remarks presented at the Conference on Scientists in the Public Interest: The Role of Professional Societies, Alta, Utah, September 1973.

page 179, Boorstin *quotation:* From David J. Boorstin, *The Image: A Guide to Pseudoevents in America* (New York: Harper & Row, 1964), p. 11.

page 179, Science for the People (SESPA): The group's policy toward the press has recently evolved to one of more accommodation to press customs; see "Science for the People: Comes the Evolution," *Science* 191 (March 12, 1976), pp. 1033–1035.

page 180, Sullivan *quotation:* From Hillier Krieghbaum, *Science and the Mass Media* (New York: New York University Press, 1967), p. 43.

page 180, "circus" and "suture" quotations: From J. Michael Crichton, "Heart Transplants and the Press," *New Republic* 158 (May 25, 1968), p. 28ff.

page 181, Abelson quotation: From Hillier Krieghbaum, *Science and the Mass Media* (cited above), p. 42; Abelson expresses similar views in Philip H. Abelson, "Science Reporting," *Science* 139 (January 18, 1963), p. 177.

page 181, "It was a big mess" quotation: From Jerry Gaston, "Secretiveness and Competition for Priority of Discovery in Physics," *Minerva* 9 (October 1971), p. 474.

page 181, speculation on Cooley: From Thomas Thompson, *Hearts* (New York: McCall, 1971).

page 182, 1972 study of science reporters: Sharon M. Friedman, "Changes in Science Writing since 1965 and Their Relation to Shifting Public Attitudes toward Science," Master's thesis, School of Journalism, Pennsylvania State University, 1974.

page 182, Shockley's visibility: Although Shockley's visibility may be increasing, surveys in 1972 and 1973 described in the author's Ph.D. dissertation (cited above) suggest that Shockley is less widely known than most of the visible scientists included in this book. Arthur Jensen of the University of California, Berkeley, and Richard Herrnstein of Harvard University are also strongly identified with the genetics-IQ-race issue.

page 184, Shockley "to me, it seems immoral" quotation: From William Shockley, "Three Moral Postulates: Truth-Concern-Death," *Presbyterian Life* 25 (February 1, 1972); quoted in John Horgan, "Dr. Shockley's Shocking Theory," *San Mateo Times Weekend*, July 1, 1972.

page 184, Shockley "It is my intention" quotation: From Leroy F. Aarons, "Intellectual Racism?" *Washington Post*, March 12, 1972, p. A-11.

page 185, King quotation: From Larry L. King, "The Traveling Carnival of Racism," *New Times* 1 (December 28, 1973), p. 40.

page 185, Rogers quotations: From Michael Rogers, "Brave New William Shockley," *Esquire* 79 (January 1973), pp. 152–153.

page 186, Shockley quotation: From Larry L. King, "The Traveling Carnival of Racism" (cited above), p. 38.

page 187, Provine quotations: From William B. Provine, "Geneticists and the Biology of Race Crossing," *Science* 182 (November 23, 1973), p. 796.

page 188, Shockley "I conducted an evaluation" quotation: From William Shockley, Commentary on *National Enquirer* article, Reprint No. 383.1 in Shockley's files, June 23, 1973. The *Enquirer* article is Jack Kelley, "Nobel Scientist Proposes Sterilization Plan to Limit Birth of Children with Low IQs," *National Enquirer*, May 20, 1973, p. 8.

page 188, "bounty hunter" story: An example is Karen Klinger, "Sterilization 'Bounty' for Low-IQ Individuals," *San Jose Mercury*, June 15, 1973, p. 1A.

page 188, coverage of Roy Wilkins story: An example is "Shockley Asks Tests of Blacks," *San Francisco Chronicle*, July 6, 1973, p. 34.

page 189, Perlman-Shockley exchange: The *San Francisco Chronicle* articles are: David Perlman, "Touchy Race Talk before San Francisco Club," January 14, 1967, p. 1ff.; William Shockley, " 'Gap' or 'Overlap' in Touchy Race Talk," January 17, 1967, p. 34; Perlman, "Race and Intelligence," January 18, 1967, p. 42; and Shockley, "The Debate on Race and Intelligence," with editor's note, "Mr. Perlman Replies," January 30, 1967, p. 40.

page 190, Leeds University story: Articles referred to are: Malcolm Stuart, "Transistor Man Gets Message," *Guardian* (London), February 15, 1973, p. 1; "Leeds Snub for Nobel Scientist," *Times* (London), February 15, 1973, p. 18.

page 191, "lemons" motto: A similar motto is the title of an autobiographical book by Warren Hinckle, *If You Have a Lemon, Make Lemonade* (New York: Bantam, 1973, 1974).

page 191, Shockley "I believe" quotation: From William Shockley, Mimeographed handout circulated at debate on IQ, Race, and Heredity, Stanford University, Stanford, Calif., January 23, 1973.

page 191, Shockley "death postulate" quotation: From William Shockley, "Three Moral Postulates: Truth-Concern-Death," *Presbyterian Life* 25 (February 1, 1972), p. 6ff.

page 193, Department of Defense study: William Shockley, "On the Statistics of Individual Variations of Productivity in Research Laboratories," *Proceedings of the Institute of Radio Engineers* 45 (March 1957), pp. 279-290.

page 194, Electronic News quotations: From Don C. Hoefler, "Silicon Valley, U.S.A." *Electronic News*, January 11, 1971, pp. 1, 4.

page 194, "thinking about thinking": An example of a publication in this area is William Shockley, "Three Great Inventions That Are Changing Our World. 1. Transistor," *Popular Science* 200 (May 1972), p. 97ff.

page 195, Mechanics textbook: William Shockley and Walter A. Gong, *Mechanics* (Columbus, Ohio: Merrill, 1966).

page 195, "hidden momentum": The first paper on the subject was William Shockley and R. P. James, "'Try Simplest Cases' Discovery of 'Hidden Momentum' Forces on 'Magnetic Currents,'" *Physical Review Letters* 18 (May 15, 1967), pp. 876-879.

page 196, Shockley "I recall" and "cult" quotations: From Shirley Thomas, *Men of Space*, vol. 4 (Philadelphia: Chilton, 1962), pp. 175, 191.

page 196, Shockley "It took me" and "He was one" quotations: From William Shockley, Letter published in *The Journal*, Forum for Contemporary History, April 23, 1973, p. 1.

page 196, Shockley "If that woman" quotation: From "Is Quality of U. S. Population Declining?" Interview with William Shockley, *U. S. News & World Report* 59 (November 22, 1965), p. 68.

page 197, U. S. News article on Denmark: "The 'Unfit': Denmark's Solution," *U. S. News & World Report* 60 (March 7, 1966), p. 74.

page 197, Shockley quotations on Denmark: From William Shockley, "Dysgenics, Geneticity, Raceology: A Challenge to the Intellectual Responsibility of Educators," *Phi Delta Kappan*, special supplement (January 1972), p. 297. In his book *The Legacy of Malthus* (New York: Knopf, 1977), Allan Chase argues that Shockley has misunderstood Denmark's sterilization laws, which are actually similar to laws in several states in the United States, and not eugenic in origin, purpose, or administration.

page 197, U. S. News quotation: From "Is Quality of U. S. Population Declining?" (cited above), p. 70.

page 198, Shockley "several dozen," "accidentally fell," and "several help-

ful" quotations: From William Shockley, Letter published in *The Journal* (cited above), p. 3.

page 198, letter to Stanford M. D. *quotation:* From Walter F. Bodmer et al., "The Issue of 'Bad Heredity,'" *Stanford M. D.*, series 5, no. 2 (October 1966), p. 41.

page 199, Brooks quotation: From William Shockley, "Possible Transfer of Metallurgical and Astronomical Approaches to the Problem of Environment versus Ethnic Heredity" (paper delivered at the National Academy of Sciences meeting, Washington, D. C., October 15, 1966, revised for mimeographed distribution October 19, 1966, p. 3.

page 199, Shockley "By campaigning" quotation: From William Shockley, Letter published in *The Journal* (cited above), p. 3.

page 199, M. I. T. professor quotation: From William Shockley, "Research Report: Research on American Lysenkoism in the National Academy of Sciences," mimeographed, August 25, 1972.

page 199, Shockley "My research" quotation: From William Shockley, "New Methodology to Reduce the Environment-Heredity Uncertainty About Dysgenics" (paper delivered at the National Academy of Sciences meeting, Houston, Tex., October 1970), p. 4.

page 199, Shockley "Perhaps nature" quotation: From William Shockley, "The Apple-of-God's-Eye Obsession," *The Humanist*, January-February 1972, p. 17.

page 200, Horgan quotation: From John Horgan, "Dr. Shockley's Shocking Theory," *San Mateo Times Weekend*, July 1, 1972, p. 2A.

page 200, Shockley "most of the attacks" quotation: From Leroy F. Aarons, "Intellectual Racism?" *Washington Post*, March 12, 1972, p. A1.

page 200, Time *quotation:* From "Free Speech?" *Time* 102 (December 3, 1973), p. 14.

THE VISIBILITY SYSTEM

page 201, Abelson quotation: From Philip H. Abelson, "Are the Tame Cats in Charge?" *Saturday Review* 49 (January 1, 1966), pp. 100-103.

page 201, Cooke quotation: From Robert Cooke, "Carl Sagan — A Man in a Turtleneck Who's Taking the Drabness Out of Science," *Boston Sunday Globe*, March 3, 1974, p. 28.

page 202, Herzog quotation: From Arthur Herzog, "Faking It," *Saturday Review of Society* 1 (April 1973), pp. 36-37.

page 203, high credibility of scientists: See, for example, Todd R. LaPorte and Daniel Metlay, "Technology Observed: Attitudes of a Wary Public," *Science* 188 (April 11, 1975), pp. 121-127.

page 203, Szent-Gyorgyi quotation: From *The Way of the Scientist*, edited by *International Science and Technology* magazine (New York: Simon & Schuster, 1966), pp. 111-128. On the role of intuition and subjectivity in the scientific discovery process, see also Stephen G. Brush, "Should the History of Science Be Rated X?" *Science* 183 (March 22, 1974), pp. 1164-1172; Arthur Koestler, *The Sleepwalkers* (New York: Macmillan, 1968); Ian Mitroff *The Subjective Side of Science* (New

York: Elsevier, 1974); William Provine, "Geneticists and the Biology of Race Crossing," *Science* 182 (November 23, 1973), pp. 790–796.

page 204, science courts: Chief proponent of a scientific court is Arthur Kantrowitz, chairman of Avco Everett Research Laboratory, Inc.; see Kantrowitz, "Controlling Technology Democratically," *American Scientist* 63 (September-October 1975), pp. 505–509; Philip M. Boffey, "Science Court: High Officials Back Test of Controversial Concept," *Science* 194 (October 8, 1976), pp. 167–169.

page 204, Ehrlich "some scientists" quotation: From Paul Ehrlich, "The Political and Social Responsibilities of Scientists" (paper delivered at the American Physical Society meeting, January 1972).

page 205, Ehrlich "in my view" quotation: From Paul Ehrlich, "The Political and Social Responsibilities of Scientists" (cited above).

page 206, Frazier quotation: From Kendrick Frazier, "Spotlight on the Visible Scientists," *Science News* 197 (June 7, 1975), p. 363.

page 207, Sullivan quotation: From Hillier Krieghbaum, *Science and the Mass Media* (New York: New York University Press, 1967), p. 43.

page 207, Cohn quotation: From Victor Cohn, "Are We Really Telling the People About Science?" *Science* 148 (May 7, 1965), p. 750.

Index
